WATER

About the Author

Julian Caldecott is an ecologist who has spent several years in senior consultancy positions with the United Nations Environment Programme focussing on environmental disaster management. His work throughout the developing world has included environmental education, ethnobiology, and sustainable ecosystem management. He is the author of *Deep Water* and the *World Atlas of Great Apes.*

WATER

Life in Every Drop

Julian Caldecott

First published in Great Britain in 2007 by
Virgin Books Ltd
Thames Wharf Studios
Rainville Road
London
W6 9HA

A catalogue record for this book is available from the
British Library.

ISBN 978-0-7535-1350-7

Typeset by TW Typesetting, Plymouth, Devon

Printed and bound in Great Britain by
Mackays of Chatham PLC

1 3 5 7 9 10 8 6 4 2

For Moyra

'The whole world has now become like one family, almost like one body. So some destruction of some other part of the world is actually destruction of yourself.'

His Holiness Tenzin Gyatso, the 14th Dalai Lama of Tibet

CONTENTS

MEASURING WATER AND LAND

1 cubic kilometre (km³) = 1 billion cubic metres (m³) = 1 billion tonnes of water

1 m³ of water = 1,000 kilos = 1 tonne of water

1 gigalitre = 1 billion litres = 1 million m³ = 1 million tonnes of water

1 square kilometre (km²) = 100 hectares (ha) = 247 acres = 0.386 square miles

1 ha = 10,000 square metres (m²) = 2.47 acres = 0.004 square mile

Foreword

As I write this, Britain is having a crash course in water awareness. Many of this book's themes are suddenly becoming all too familiar. We have realised, for example, that our vulnerability to flooding is made worse by building on floodplains, by channelling rivers through narrow, artificial banks, and by covering ground with tarmac and concrete, thus preventing it from absorbing water. We've learned the irreplaceable value of clean drinking water, and we are coming to realise that climate change is steadily demolishing our expectations about a gentle English climate.

Water is an extraordinary substance that makes life on Earth possible. But almost all the world's water is salty, and for us on land a regular supply of fresh, clean water is uniquely precious. Such a supply is the most important service that an ecosystem can offer, yet is often allowed to lapse through abuse, or it is diverted from those who really need it, or destroyed by over-use or pollution. As a result, over a billion people now have no access to clean water and 2.6 billion have no effective sanitation system. One consequence is a huge waste of human energy in an endless quest for water, a burden that often falls hardest on women and children. Another is unnecessary illness, which every year claims the lives of nearly two million children. We are, in every sense, facing a global water crisis.

Yet, as Julian Caldecott explains here, this global crisis is in fact made up of tens of thousands of *local* water crises, each one due to decisions that affect local ecosystems. Water and ecosystems are linked, from the boundaries of each catchment to the streams, rivers, lakes, floodplains, swamps and estuaries created by water. Everything we do in a catchment affects what happens downstream, so logging, farming and grazing, applying fertilisers or pesticides, dumping garbage, releasing sewage or spilling chemical wastes all have an impact that's conveyed by the ultimate solvent, water. Meanwhile, we've taken to pumping water from the ground at rates far higher than it's being replaced, causing wells to run dry in region after region.

This book is about human decisions. Julian Caldecott draws stories from all over the world, and reveals the many ways in which our experience of water is a common one. He shows how different approaches can have different outcomes, some destroying ecosystems, some transforming them, some sustaining them. But there's also a bigger context. The viability of the biosphere depends fundamentally on water – often in ways that we barely understand. Water, ecosystems and climate are inextricably linked, so we need to make wise decisions about all three. Local ecosystems determine whether or not there's water in your well, river or tap, and help dictate rain or drought, storm or famine. All the evidence is that the negative changes we make to our environment are contributing to mass extinction, local water crises, and further climate chaos.

This fascinating book explains not only why we need to restore balance, but more importantly how we can do it.

Zac Goldsmith
Director, *The Ecologist*
London, July 2007

1. WHAT *IS* WATER?

We think we know water. We take it in through almost everything we drink or eat. We wash in it, swim in it, float on it. We wish it would stop falling on our heads, or hope it will start. We know that we have to drink several litres of it each day, depending on heat and sweat, for if we don't we are first driven and then tormented by thirst. We know that there's water in everything that comes out of our bodies, from blood and breath to tears and spit. We know that we must water our plants if the rain will not, and that we must bring our pets and livestock to water every day. We know that things that are plump and moist are usually alive, and that things that are shrivelled and dry are usually dead, or waiting for water to make them good again. We know that rivers flow downhill, and that there are fish in them, and otters or platypuses near them. We know that the ocean is vast and powerful, and that there are whales and sharks in it. We know that water is potent and symbolic, for we dab it on our heads or sprinkle it on the ground during rituals.

We know a lot about water without really thinking about it. We have an intuitive appreciation of its key role in physiology (because of thirst and sweat), ecology (because of fish and farming), magic (because of rituals and dreams), economics (because we often have to buy it) and power (because we are vulnerable to those who control it). But this everyday experience is just a fraction of the total reality of water and our relationship with it, our collective struggle to understand and use it in all its dimensions. This book explores some of these aspects of water, as seen through the prism of ecology. I chose that particular prism because life, water and ecosystems go together, always. We are alive, and in using water and ecosystems we determine what all life will be like in the future, including our own.

In Chapter 2, we'll find out about the role of water in the biosphere, the 30 km-deep living skin of the Earth. Here we'll look at the origin and evolution of life in symbiosis with water, the accumulation of biodiversity over evolutionary time, the ferocious culls of mass extinctions, the threat of global warming, and the role of the Earth itself in correcting it if we don't. Chapter 3 will describe our experience of water in evolution, and its significance for how we think about and try to solve environmental challenges, including the water crisis. Other chapters will focus on the main water-bearing ecosystems of our planet: the oceans, wetlands and swamp forests, the lakes and rivers, the ground waters and aquifers, and the farms that they sustain. In each case we'll look at the natural history of the ecosystems themselves, as well as how people have used and abused the living things and waters within them.

Each chapter will show how a particular kind of thinking leads people to impose short-term demands on nature, with disastrous consequences, and how another, more ecological kind of thinking leads us to more sustainable outcomes. It is the dominance of short-term thinking that has led us to environmental crisis, the solutions to which may be found in re-discovering the other kind of thinking. The last two chapters explore this possibility. Chapter 9 will examine the international efforts we've made so far to conserve nature and water, while in Chapter 10 we'll look at what we can do as societies and individuals to restore the biosphere to harmony, how we've solved similar major problems in the past, and we'll see a vision of what the biosphere might look like in the year n2085 if all goes to plan. The water crisis is deeply challenging, but we can approach it knowing that it, like many of the world's problems, involves *local* ecosystems and *local* communities, so although they are all connected, we do have the power and the precedent to solve these problems bit by bit.

But we'll start here with where water came from. We'll then look at its physical and chemical nature and behaviour, to give a raw insight into the properties that make water so important and also so strange. And as we go, we'll see how each of the extraordinary properties of water contributes to a symbiotic relationship with life itself, from the innermost workings of every living cell, to the physiology of whole organisms and the patterns that sustain ecosystems and, ultimately, the biosphere itself.

ULTIMATE ORIGINS

Water's been here on Earth for a long time, and its peculiar properties have always been important to the planet's evolution. But where did the stuff come from, and how did it get here in such vast and fortunate quantities? In the big picture, water ought to be common, since its molecules are made of atoms of two of the commonest elements of ordinary matter in the Universe: hydrogen, which at 75 per cent of everything is the most abundant element, and oxygen, which at far below 1 per cent is still the third most common (after helium, at 23 per cent).

Hydrogen condensed out of the primal chaos of the Big Bang, a few seconds after the beginning of the Universe. But oxygen and all the other elements with larger and heavier atoms were created much later, in complex thermonuclear reactions deep within stars. When these stars eventually died, the larger ones exploding as supernovas like SN 2006gy, the brightest ever seen, all the elements within them were scattered deep into space. There they accumulated through the slow action of gravity into new generations of stars, as well as into the spinning clouds that would one day become planets and all the other solid forms out there, such as asteroids and comets.

Some of these clouds happened to contain more water than others, and some, after collapsing to become solid, happened to have the right conditions to keep it. The strong gravity field of a massive planet would hold it down, and if there were little enough radiation from nearby stars the water wouldn't be broken back into separate hydrogen and oxygen atoms. If there *was* sufficient radiation, the lighter hydrogen atoms would mostly drift off into space, leaving oxygen to combine with something else, and so water would have gradually vanished. A low enough temperature would keep water on the solid planet too, as deep-frozen water mixed with rock is immobile until it gets heated up.

Water has so far been found on three planets: Earth and Mars in our own solar system, and on HD 209458b, a Jupiter-like gas giant located 150 light-years away in the constellation Pegasus (we know this because of the way light is absorbed in the wavelength characteristic of water vapour as it passes across its star's face). The presence of hot, solid water under great pressure has also been deduced from the density of a fourth planet, GJ 436b, which orbits a cool, red star some 30 light-years away. Water has also been discovered or confidently inferred on or inside several moons, including our own, Jupiter's Europa, Ganymede and Callisto, and Saturn's Enceladus and Titan, as well as in several comets.

Moreover, from its light-spectrum signature, water vapour has been detected in the hot atmosphere of our own Sun, and in clouds of inter-stellar gas, both within our own galaxy (such as in the Orion Molecular Cloud, 1,500 light-years away), and in other distant ones (such as the spiral starburst galaxy NGC 253, the elliptical galaxy NGC 1052, and the Circinus galaxy ESO 97-G13). This is a tiny sample of the Universe, but already the hint is that water is very widespread, and we can expect to find it quite often as our search technologies improve.

COMET WATER, EARTH WATER

In 2005, the US spacecraft *Deep Impact* launched a 370-kg payload into a head-on collision course with the rocky nucleus of the small comet Tempel 1, and retired to a safe distance. The combined speed of the two objects was about 37,000 km/hour, so a considerable explosion was anticipated. What was not expected was that about a quarter of a million tonnes of water would be blasted from the nucleus, which was not believed to be very icy, and continued to leak out over thirteen days.

It is thought that a more typical comet nucleus fits the description of a 'dirty snowball', with Halley's Comet, for example, having a mass of about 100 billion tonnes, most of it ice, and others being up to ten times as big. A million comets of this size hitting the Earth would have gone a long way to filling the seas. Since comets were common in the inner solar system early in Earth's history, it's possible that the planet was hit by a large comet once every thousand years for its first billion years, which would have done the trick.

Not all comets, however, contain water that has the same isotopic composition as water found on Earth, i.e. the same ratio between ordinary water and 'heavy' water. The latter is water in which the hydrogen atoms are replaced by deuterium (a stable *isotope* of hydrogen with an extra neutron in its atomic nucleus). One that did, though, was Comet Linear, which broke apart near the Sun in 2000, yielding a cloud of hydrogen formed by the disassociation of an estimated 3.3 million tonnes of Earth-type water.

It seems likely, therefore, that at least some of Earth's water arrived when the planet was bombarded by comets early in its history. But there is also evidence for an additional mechanism, in which the Earth's atmosphere, even today, is being bombarded by small comets made of pure water. Ultraviolet satellite images of Earth's atmosphere show what look like very-high-altitude holes or vapour trails, hundreds of them appearing each day. These images have been studied since the mid-1980s by a team led by Louis Frank, Professor of Physics at the University of Iowa and a leading authority on energetic charged particles, plasmas and auroral imaging around the Earth and elsewhere in the solar system. Frank and his co-workers interpreted the images as showing the breaking up of small comets. These they visualised as loosely packed 'snowballs', 20–40 tonnes in weight, that disintegrate from rapid electrostatic erosion as they approach the Earth and are then vaporised in the upper atmosphere.

Objects of this size are nearly invisible in space, especially if they are coated with dark dust, so direct observations of these comets are unlikely. Further satellite evidence for them was obtained in the late 1990s, however, including signs of water being released at between 960 and 24,000 km in altitude, and a photograph of what could have been the trail of a small comet vaporising over the Atlantic Ocean at between 8,000 and 24,000 km in height. If the interpretations are correct, which is strongly debated, then we're looking at the arrival of a small water body weighing about 30 tonnes every few seconds – a rate that if sustained could deliver a metre or two of depth to the world's oceans in a million years.

So all in all we've a lot to thank the Universe for, and much to wonder about. We're made of elements forged within stars and blasted across space by supernovas. One of those elements teamed up with the commonest substance in the Universe to make a compound that is, as we'll see, both unique and perfect for sustaining life. And wandering comets brought (or maybe are still bringing) enough of it here for the Earth's particular gravity and distance from the sun to allow the creation of a blue planet.

MOLECULAR BONDING

Water is a truly remarkable substance, with properties like no other. These properties, and the transformations of water from icy solid to liquid, from its liquid state to vapour, and back again, are central to life on our planet. Water's unique attributes result from the forces at work within and between its molecules, and understanding them is a vital key to making sense of nature. So it's important to grasp some basics of chemistry – the shape of water molecules, and how they behave in partnership with each other and with other substances – in order to understand why it's so important for life on Earth.

All chemical matter, including water, consists of elements, either as pure masses of one type of atom, or else, more often, as masses of molecules, which are made of joined atoms. If more than one kind of atom is in a molecule, it's called a *compound*. In a molecule, the atoms are strongly *bonded* to one another, by arrangements based on the rules that opposite charges attract while similar charges repel one another. Since protons in an atom's nucleus are positively charged, and the atom's electrons orbiting the nucleus are negatively charged, there is scope for sharing or transferring electrons between atoms that approach each other. In a *covalent bond*, a pair (usually) of electrons is attracted into the space between the two atomic nuclei. Once there, the electrons and nuclei are pulled together. But the two positively charged nuclei also repel each other, so the two atoms stay at a distance where the attractive force balances the repulsive force: a point of equilibrium.

Although the number of electrons and protons in an atom are usually the same, so the atom as a whole is uncharged, this is not always so. An atom with a charge (either positive or negative) is called an *ion*, and these are at the root of the other main kind of bond between atoms, the *ionic bond*. Ions can be created when an electron is lost, for instance due to the impact of radiation, or when a strongly charged nucleus approaches a more-weakly charged one. In this case, the positive charge of one nucleus overwhelms the positive charge of the other, forcing a transfer of electrons to balance things out, and both become ions. Since one ion is positive and the other negative, the atoms are pulled together. However, as the atoms approach their electrons are drawn into the space between their nuclei and they create covalent bonds as well. Ionic bonds often form between metallic and non-metallic elements – an example being table salt, which is a compound of sodium (a metal) and chlorine (a non-metal).

The details of how and why atoms combine to form molecules, and how molecules interact with one another and otherwise behave, is the subject of chemistry, and some of its discoveries are needed to explain the properties of water. For water is a compound, its molecules comprising one atom of oxygen and two of hydrogen, which is why its chemical formula is H_2O. These atoms have covalent bonds between them, with one hydrogen atom on either side of the oxygen atom. Each hydrogen atom brings with it one orbiting electron, and each oxygen atom brings six. In the covalent bond of the water molecule, the single electron of the hydrogen atom is paired with one of the six electrons of the oxygen atom.

With two such bonds in each molecule, there are two hybrid pairs of electrons and two pairs of electrons that are not involved in the bonding. Thus the oxygen atom is surrounded by four electron pairs, all negatively charged. Since they repel each other, the pairs arrange themselves as far from each other as possible. All else being equal, this would create a four-pointed structure or tetrahedron, with a regular angle between each of the participants of 109°. But instead, because the two non-bonding electron pairs remain closer to the oxygen atom's nucleus, these more strongly repel the two bonding pairs, thus

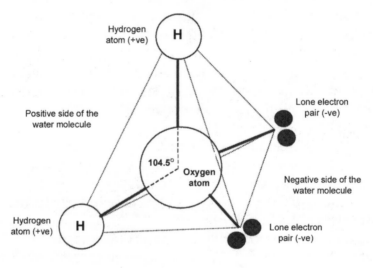

The distorted tetrahedral shape of a water molecule

pushing the two hydrogen atoms closer together. The result is a distorted tetrahedral arrangement in which the hydrogen-oxygen bond angle is only 104.5° (see below). Remember this diagram: it might be the most important shape you'll ever see, for in it is contained the power to bond, to dissolve, to shape, to convey, and to transform.

AT THE SIGN OF THE TWISTED TETRAHEDRON

Each water molecule, then, is an oxygen atom surrounded by four points, two of them, on one side, being hydrogen nuclei, and the other two, on the other side, being pairs of electrons. The hydrogen nuclei have a positive charge, since their electrons are closer to the oxygen nucleus than they are to their own hydrogen nucleus, while the electron pairs have a negative charge. This arrangement has two important, but linked, consequences. First, one side of the molecule is positively charged and the other negatively charged, making the water molecule *polar*. This means that the positive side of a water molecule is weakly attracted to the negative sides of other water molecules, and vice versa, and that they are also attracted to their opposites in any other polar molecule that they meet.

Not all molecules are polar but many are, because they have a net positive and a net negative part, side or end. They include molecules with a hydrogen-oxygen group at one end (for instance some sugars, like glucose, and alcohols, such as the ethanol in alcoholic drinks), and molecules with a hydrogen, oxygen or nitrogen atom at one end (like water itself, and ammonia, and many biomolecules or parts of them). All of these have a mutual attraction to water, so are called *hydrophilic*, and the smaller ones all dissolve in water. The same applies to most compounds with ionic bonds, since the negative–positive attraction that holds them together can be switched to water as they dissolve in it. Shapes of stable, highly structured water molecules are created around each polar molecule or ion. These are called *hydration shells*, and are like moulds or negative images of the substance that water has encountered. Some believe that these hydration shells can persist even after the substance that made them has departed, suggesting that water may have some kind of 'memory'.

One result of its polarity is that water is an excellent solvent, so it can pick up and carry many other kinds of substance. Another is that it can help organise complex biological molecules, for instance in a cell, by attracting those parts of large molecules that are hydrophilic, and repelling other parts that are non-polar and hence *hydrophobic*. Cell membranes, for example, are made of layers of phospholipids (phosphorus compounds attached to fats or oils), and in a watery environment the hydrophobic oily ends hide away inside the membranes, making the whole structure possible. So too, all-important protein molecules often depend on their shape to have their proper function, and rely on the presence of water to guide their hydrophilic and hydrophobic sections into the right places. Without water, these biomolecules would unravel and cease to function, and many require bondings with very specific directionality which only water can provide.

The watery environment can hold many molecules and ultrafine particles in suspension, forming what are called *colloids*, with their own unique properties. Notably, such water-based colloids can easily be transformed from a fluid or *sol* state into a semi-solid *gel* state, and back again: transitions that underlie many cellular mechanisms. As Philip Ball put it in his book *H₂O: A Biography of Water*: 'That the only solvent with the refinement needed for nature's most intimate machinations happens to be the one that covers two thirds of our planet is surely something to take away and marvel at.'

The other consequence of the twisted tetrahedron is that each water molecule is able to form four *hydrogen bonds* with other molecules, positive (water's two hydrogen atoms) to negative parts of other molecules, and negative (water's two lone electron pairs) to positive parts of other molecules. These hydrogen bonds (H-bonds) are about ten times stronger than the forces of attraction between polar molecules, but about ten times weaker than the covalent bonds between atoms. They are, however, strong enough to make water molecules decidedly 'sticky', and it is these hydrogen bonds that dominate the structure of water. In liquid water, then, every molecule spends most of its time H-bonded in all directions.

This is not a static arrangement, though, as the molecules switch H-bonds very rapidly, in trillionths of seconds. So the whole thing is both highly structured, being a single, huge H-bonded cluster, and also highly dynamic, since H-bonds break and re-form with lightning speed. There's also some kind of co-operation involved, because the forming or breaking of one H-bond alters the chance that another will be made or broken nearby. Why or how this happens is one of the mysteries that surround water, while at a practical level making it very hard to describe or model the structure of liquid water.

Although water molecules are unique in being able to form as many as four H-bonds each, this form of bonding is not exclusive to water. Any molecule that has a hydrogen atom attached to an oxygen or a nitrogen atom is capable of H-bonding. This includes alcohols such as butanol and ethanol (which are also polar molecules), which contain groups of oxygen and hydrogen atoms, and other carbon-based (or *organic*) molecules that have groups of nitrogen and hydrogen atoms. These range from simple molecules like methylamine to large ones like proteins and DNA. Hydrogen bonds help biological molecules form and maintain their proper shapes, including the DNA double helix, the two strands of which are held together by H-bonds between hydrogen atoms attached to nitrogen atoms on one strand, and lone electron pairs on another nitrogen or oxygen atom on the other strand. The net effect is that a firm but breakable chain of H-bonds links the helices together, until they need to be 'unzipped' to allow them to be read into RNA or copied into DNA, thus allowing the chemistry of life and heredity to occur.

THE EFFECTS OF WATER'S H-BONDS

As heat energy is fed into a liquid, its molecules bounce around with increasing vigour. Eventually they vibrate so hard and fast that they start losing contact with one another, and the liquid boils away into a vapour or gas. This *boiling point* depends on pressure as well as temperature, since high pressure holds the molecules together in a liquid form up to a higher temperature than at low pressure. This is why 'the' boiling point of water is always described as 100°C at sea level, i.e. under one Earth atmosphere of pressure. If you increase this, for instance by heating water in a pressure-cooker or autoclave, it will be *superheated* to much more than 100°C before boiling. But if you heat water at a high altitude, under low pressure, it will boil at a much lower temperature than 100°C, making it hard to make a decent cup of tea.

Pressure is not the only thing that holds molecules together, though. Another is hydrogen bonding. The boiling point of a liquid ought to be roughly related to the size of its molecules, with small ones bouncing around more at lower temperatures than large ones because of their lower mass and weaker bonding forces. By comparison with other liquids made of molecules of about the same size, such as methane and hydrogen sulphide, water would be expected to boil at about *minus* 90°C, not *plus* 100. Other liquids with H-bonding, such as hydrogen fluoride, ammonia and ethanol, show the same anomalously high boiling points, though not as strongly. Without H-bonds, water would exist in our world solely as a vapour, creating a grossly amplified greenhouse atmosphere and a surface temperature of hundreds of degrees, rather like Venus where the air is hot enough to melt lead. But instead, on Earth, water has existed as a liquid for almost all the time and in almost all places over the last several billion years.

The temperature rise in a substance caused by a certain amount of heat energy being put into it is called its *heat capacity*. This is extremely high in water, because of its structure, so more heat is needed to raise its temperature than almost any other substance, and more heat has to be lost to cool it down. This means that blood can carry heat away easily from working muscles and other hot organs, helping to keep the whole body at an even temperature of 37°C. It also means that ocean currents can carry phenomenal amounts of heat, ensuring that the world as a whole is kept at a relatively constant temperature. Without water's high heat capacity, the Earth may well have been uninhabitable, except possibly in patches.

Related to its heat capacity, water also has a very high *latent heat*, which is the heat energy absorbed by a substance as it changes from liquid to vapour, or released when it changes back again. Thus, lots of heat is taken up when water evaporates on the skin, making sweating an effective way to cool the body down. Water also *conducts* heat unusually quickly, again helping bodies to stay evenly hot, although it can make scuba diving in cold water very chilly, and diving near volcanic vents very dangerous.

Liquid water has unusually strong *surface tension*, since the H-bonds hold the surface molecules together in a skin, upon which all manner of small animals can run (like water skaters) or dangle (like mosquito larvae). High surface tension also allows *capillary action*, by which water creeps upwards in small spaces, for example raising underground water through the soil to the roots of plants, and then internally upwards to their leaves. By tugging at solid surfaces, it also helps erode rocks into tiny particles of silt, from which chemicals can more easily be dissolved, thus helping to create soils, and bearing nutrients through ecosystems.

IT'S ONLY PHYSICAL – EXPANSION AND CONTRACTION

Like other substances, liquid water contracts and becomes denser as it cools. Cold surface water encounters air and dissolves oxygen, so by sinking it delivers oxygen to deep water, where it supports life that would otherwise suffocate. Also like other substances, water expands as it's warmed, but while most liquids do all this expanding and contracting from the moment they melt, fresh water only does it from 4°C. From that point it expands whether you heat it or cool it. The cooling expansion is due to the H-bonded molecules forming large clusters, which push them apart. But the system is transformed when it freezes, since the H-bonded molecules suddenly enter a new hexagonal lattice formation, which is about 9 per cent bulkier than cold liquid water.

A 9 per cent expansion on freezing makes water truly unique, since other substances become denser when they freeze, and the fact that ice floats on liquid water has awesome consequences for life. Taken together with the density maximum at 4°C, which makes cold water sink, it means that the cooling of a fresh water body such as a lake or pond isn't just a surface event – the whole thing has to be around 4° before *any* freezing can happen. Then freezing only occurs from the top down, and usually stops when there's a layer of ice floating on the surface. Hence the deeper waters remain liquid, at 4°C, and fish and other organisms can survive there throughout the winter. This is slightly different for sea water, since its salt content lowers the freezing point by about 2°C, and also lowers the temperature of the density maximum. The effect of this on sea water is to make the deep waters of cold oceans like the Arctic about 4°C colder than the depths of frozen-over fresh waters. This is a physiological challenge for Arctic fish, many of which have antifreeze molecules in their blood to cope with zero-degree water. But ice does float on both sea water and fresh water, which is fortunate since otherwise ice would sink like stone, filling up the bottoms of the oceans and lakes far from the warming sunlight of spring. In other words, the world's water bodies would quickly become solid ice, with a seasonal layer of liquid water on top.

ANOMALIES AND ICES

The list of other water anomalies is a long one, over sixty at the last count, and all are probably connected one way or another to its exceptional H-bonded structure. Strangenesses include those to do with temperature, with cold liquid water shrinking as it's heated, becoming harder to compress, easier to heat, less able to dissolve gases, slowing light more and sound less, and hot water doing the exact opposite on all counts. Meanwhile, with increasing pressure, cold water molecules move faster and the water becomes more runny, but hot water molecules move more slowly and the water becomes more viscous. Finally, no other material is commonly found as solid, liquid and gas, with local and seasonal transitions between these forms driving most if not all of the world's ecology.

Did I say *finally*? I meant, until we look at ice. This, it turns out, comes in a dozen different kinds, depending on pressure. Regular ice at low pressure, i.e. from below one *atmosphere* up to a thousand or so, is made of hexagonal structures of H-bonded molecules, with plenty of space inside. But increase the pressure and you start to see the molecules slip suddenly into new configurations. Two new forms happen between 1,000 and 3,500 atmospheres, and another three (one of which can only fleetingly exist) between 3,500 and 20,000 atmospheres. The first four of these are all more-or-less deformed versions of the regular ice H-bonding matrix, but the fifth, and two others that form above 20,000 atmospheres of pressure, are organised in interlocking networks, packing more than twice the number of molecules into the same space, while at even more extreme pressures the hydrogen bonds themselves are completely reinvented. Not surprisingly, these various forms of ice behave in odd ways that physicists find fun, with one form having a melting point of 80°C and another of over 100°C, but only if they are kept under pressure. This is fortunate for life, since water that froze only at that kind of temperature would be lethal if it got out of the laboratory and spread.

There are interesting things to find at very low temperatures too, even outside super-pressure chambers. If you condense water vapour quickly on an ultracold surface, you get a kind of fluid ice that's as viscous as molten glass, which survives from minus 120°C to minus 140°C. But then, if you put regular ice under 10,000 atmospheres of pressure at minus 196°C you get another kind of glasslike ice, and at several thousand atmospheres and temperatures below minus 75°C, there may be two kinds of liquid water as well. This all seems to show water seeking the best possible configuration and bonding pattern among its molecules to reconcile contradictions in the energy environment that experimenters have created. This sort of responsiveness and adaptation is almost lifelike, and makes one wonder anew about what water actually is.

VIBRATIONS IN THE LATTICE

Returning to regular ice, the mysteries aren't over yet as we haven't looked at snowflakes. These little crystals are famous for their unique shapes, many of them very beautiful. They form in certain conditions of moist, cold air, but in other conditions the ice can make plates, needles, prisms and other shapes instead. The classic snowflake grows identical fern-like tendrils along paths that diverge at 60° because of the hexagonal arrangement of molecules in ice. They show an extraordinary, inexhaustible creativity of form, and no other substance crystallises in so many different shapes. But the big question is why are snowflakes symmetrical, with every branch identical to every other? It's as if each of the six branches somehow knows what the others are doing, and does the same. There are two main hypotheses to explain this. One is that all the branches are similar because they happen to grow close together under the same conditions at the same rate. The other is that there are vibrations in the crystal lattice of the growing ice, which bounce back and forth through the crystal, thus organising it. The first idea sounds weak, while the second begs the question of what kind of vibration in what kind of field could achieve this? Rupert Sheldrake, an eminent biologist, neatly incorporates snowflake symmetry into his theory of morphic resonance, envisioning morphogenetic fields that organise growth and form in everything. Maybe so, but that's another story.

DIFFERENT WATERS

By now it should be clear that water is a simple thing, but also very complicated. And there's one final, *final* thing. When we drink water, we sense that there's something different between a scoop from a mountain spring and a mouthful taken from a bottle of well water, or a glass fresh from the tap and one that's been standing overnight. There are all sorts of ways in which these waters may differ, including all manner of dissolved chemicals, but there's also a feeling of energy and liveliness that varies. Fresh spring water has a buzz to it that dead river water doesn't possess. This is hard for chemists to analyse, but people feel it anyway.

Water is inherently dynamic: it stores and responds to energy. We've seen how the great complexes of water molecules seethe with breaking and re-forming hydrogen bonds, billions in every instant, and all this activity alters radically with temperature and dissolved salts and gases as the ever-changing labyrinth of molecular bonds responds to minute signals. Many of the molecules in liquid water are fragmented into hydrogen (H^+) ions and hydroxide (OH^-) ions, and the proportion can vary hugely with light, heat and dissolved materials. Adding even a little acid increases the number of hydrogen ions, while adding any kind of base increases the number of hydroxide ions. Meanwhile, hydrogen ions can 'move' almost instantly along chains of linked water molecules – if one is added at one end of a chain, another appears at the other end, as if by magic. By putting a sample of this extraordinary substance in our mouths, it's hardly surprising that we can sense its condition, somewhere in the vast range of energy states that water can adopt, and feel the way it instantly changes in response to new conditions even as it touches our lips.

2. WATER IN THE BIOSPHERE

THE BIOSPHERE

The biosphere is the name we give to all the parts of the Earth where life occurs, the maximum extent of all ecosystems. Water exists throughout the biosphere, inside living organisms, and as ice, vapour or liquid throughout their environments. There are about 1.4 billion cubic kilometres (km^3) of water on Earth, of which almost all is sea water. Of the 36 million km^3 of fresh water, two-thirds is frozen in ice caps and glaciers, and only about 12 million km^3 is liquid, almost all of which is held underground in rocky aquifers. The remaining fraction of about 200,000 km^3 is found above ground, or nearly so. Of this, some 90,000 km^3 is in lakes, 90,000 km^3 is in soils and permafrost, 13,000 km^3 is atmospheric water vapour, 11,000 km^3 is swamp water, 2,000 km^3 is in rivers, and 1,000 km^3 is contained inside living organisms.

All these figures are approximate, and prone to change as new discoveries are made – such as on the size of aquifers and the amount of water bound into deep rocks – and as climate changes increase the melting of permafrost, ice caps and glaciers. The general pattern of water in the biosphere, though, is that it moves through two kinds of cycle. There is a slow cycle, in which water is held for millennia in aquifers and in long-term ice, only gradually seeping onto the surface through springs, or streams from melting glaciers. Then there is a fast cycle, in which about 500,000 km^3 of water evaporates every year from sea and land, becoming vapour that condenses to clouds and falls as rain. It is this fast cycle that irrigates the dynamic parts of the biosphere on land. The other parts, the spores and seeds trapped in long-term ice, are best described as still life.

The largest volume of the biosphere is gas, our atmosphere, which is inhabited to great heights by tiny spiders drifting on silken threads, by the spores of bacteria, fungi, ferns, lichens, mosses and other organisms, and by the lightest of wind-dispersed seeds and pollen grains. These push the upper limit of the biosphere up to an altitude of at least 17 km, which is about the extent of the upper limit of the *troposphere* at the equator. This inner layer of the atmosphere is constantly mixed by rising air, but at its top it gives way to the jet streams and the stratosphere above them. At lower levels, we find birds typically flying from near ground level to a height of about 2,000 metres, but some go much higher. Bar-headed geese migrate over the Himalayas at 8,300 metres, and have been seen above that. In 1973, a Ruppell's griffon vulture collided with a commercial aircraft over Abidjan, Côte d'Ivoire, at a height of 11,280 metres. At such an altitude, more than 90 per cent of the mass of the atmosphere lies below the bird's wings, and temperatures are around minus 50°C. Such high-flying birds, which also include turkey vultures, buzzards, ospreys, terns and cranes, occasionally go into the upper troposphere and ride jet streams there.

Life becomes far denser within the biosphere close to, on, and just beneath the Earth's surface, where all its key resources come together. These include the sunlight and warm dense gases of the lower atmosphere, which may also be dissolved in water to sustain aquatic life. In particular, there is oxygen to support aerobic respiration, and carbon dioxide for photosynthesis. There is also nitrogen, the air's commonest ingredient, which is fixed as nitrates by soil and root bacteria.

The resources of the planet's surface also include the trace nutrients and minerals needed by plants and animals. Examples include magnesium, a component of chlorophyll which traps sunlight in plants, and iron, part of the haemoglobin molecule which transports oxygen in mammals. Thus, on land there is the familiar bustle of vertebrate and invertebrate animals and plants, the burrows and roots of which penetrate tens of metres into the soil. This life draws on resources of light, air and soil, but largely feeds on itself, obtaining most of the carbon, nitrogen, phosphorus and other elements it needs by recycling the dead and the wastes of the living.

In the sea, larger life forms dwell far deeper than on land, with fish and invertebrates thriving in the abyssal depths and deep ocean trenches close to 11 km below the surface. All is ultimately sustained by the largest photosynthetic mechanism on Earth, the phytoplankton of the top hundred metres of sunlit water. Everything else is supported by recycling – the capture of nutrients from dead creatures falling from above – or the blooming of life that happens when sea-bed sediments are up-welled into surface waters. But both on land and at sea, microbial life penetrates far deeper still. There are probably bacteria living kilometres down in the sediments beneath Lake Baikal in Russia, and under the deep sea bed, and they and their spores are carried by water seeping into many deep places of the world. They must push the biosphere's inner limits nearly as far down as it extends upwards, making the whole thing at least 30 km deep, from cold space to hot rock.

THE HYDROSPHERE

For most purposes, the biosphere is the same as the hydrosphere, where water is found, since water and life are so intimately connected. But there is a difference, and not a subtle one either, as it involves unimaginable temperatures and pressures. For the biosphere and hydrosphere come apart deep underground. We'll see in Chapter 8 that much of the Earth's crust is continually being recycled, as huge chunks of the surface are driven under each other, forced down into the fiery depths of the planet. The living creatures dragged down by such submerging plates can't survive when the water in their very molecules is being boiled away. The water, though, continues downwards, so the hydrosphere must be much larger than the biosphere.

There is also the possibility that large amounts of water were incorporated within the deep levels of the Earth from its beginning, so not all the planet's water exists on or near its surface. There are signs that water is bound into rocks at great depth, even as deep as 400–500 km, where the temperature exceeds 1,000°C. If so, the amount of water stored down there would be many times greater than that in all the seas of the Earth, and the hydrosphere many times more voluminous than the biosphere. There are also signs that some of this deep water may find its way into near-surface locations, to participate in volcanic eruptions, and even perhaps to recharge certain aquifers. If so, then water would be providing a more direct link than we'd thought between the deep interior of the Earth and the near-surface biosphere.

That said, our focus here is on the water–life combine, an ancient symbiosis that depends on the extraordinary properties of both water and life. This is the biosphere, and this chapter explains how it developed, how it works, and what's been going on in it recently. The last point is important, as well as scary, since there are ample signs that the living world is becoming dangerously unstable. What looks like a major transformation in all the systems of the biosphere seems to be underway. As we'll see, many people think of this as a restoration of a state of dynamic equilibrium that has prevailed for ... well, who knows? The biosphere's always changing, and life has always adapted one way or another, as well as contributing to change itself. The problem for us is that we are now the ones disturbing the equilibrium, and all our problems, including the global water crisis, are a cause or a consequence of this, or both.

Perhaps the significance of this statement will be clearer once we've established a few basic facts, for example about our place in nature. The first great and terrible fact is that we have only one biosphere to live in, or to experiment on. There are no grounds for thinking that others exist, or that we could get to them if they did. We, just like every other species, are stuck here on Earth. We live with each other's wastes and needs, and the consequences of our own and everyone else's actions. There's no way out of an Earth-bound life but in imagination for our minds, and in death for our bodies. That being the case, it's hard to think of a better use of time than understanding how the world works, and learning how to live on it as if we intended to stay.

BEGINNINGS

The Earth formed from a cloud of space debris that pulled itself together by its own gravity about 4.6 billion years ago. As gravitation depends on mass, heavier molecules, especially of metals like iron, tended to accumulate in the middle of the ball, and some of these were radioactive, making the interior of the Earth hot. By about 4.4 billion years ago, the Earth's surface had cooled enough to form a crust. This was heavily pierced by active volcanoes, and under fierce bombardment by extraterrestrial bodies. Meanwhile, the original atmosphere of hydrogen and helium had largely boiled away, these light gases escaping into space. They were replaced by heavier volcanic and impact gases, such as water vapour, carbon dioxide, carbon monoxide, sulphur dioxide and ammonia (later broken down by sunlight into nitrogen and hydrogen), thus creating a second atmosphere.

Within a couple of hundred million years, the system became cool enough for water vapour to condense and fall as rain. For a while, this would have boiled back off the hot surface, transferring heat into the atmosphere and thence into space, condensing, and being dragged back to the surface by gravity. Eventually, by about four billion years ago, the surface had cooled enough for the first oceans to form, under a nitrogen and carbon dioxide atmosphere. This air contained about a hundred times as much gas as our own, but as it cooled much of the carbon dioxide was dissolved in the new seas and precipitated as carbonate rocks. Shortly afterwards, about 3.8 billion years ago, life arose, mysteriously, in those seas.

All living things on Earth are descended in an unbroken line from these first, very simple creatures, which are thought to have been similar to modern archaean microbes and bacteria. Since all life uses some variant of the code-bearing, replicating molecule DNA to transmit design information down the generations, we might as well say that DNA-like replicating molecules arose on Earth, mysteriously, about 3.8 billion years ago. This could have happened, somehow, on Earth itself, but the harsh physical conditions of the planet, less than a billion years after it formed, make it an unlikely environment for an extremely complex molecule of this sort to arise from scratch. So it seems just as likely that a DNA-type molecule arrived from elsewhere, perhaps carried by the same icy comets that helped to make the Earth such a wet planet, or possibly from Mars by way of fragments blasted off its own surface by asteroid impacts. Mars features because its surface cooled hundreds of millions of years before Earth's, and fragments of rock apparently of Martian origin have been found on Earth.

In any case, early on, there was virtually no oxygen in the Earth's atmosphere, and the first organisms took their energy from chemical reactions such as *methanogenesis*. They began to contribute new gases to the atmosphere, but not oxygen, which was probably toxic to them. The lifestyle of these earliest organisms is represented in the modern world by the archaeans, a class of microbes found only in rare, harsh environments such as extremely salty ponds, oxygenless sediments and deep-sea volcanic vents. Photosynthesis may well have arisen very early, with organisms using sunlight to make carbon dioxide and water molecules combine, thus creating simple sugars and oxygen as a waste product. But the oxygen given out was at first absorbed by reactive gases and minerals. Thus, vast strata of red rocks, rich in oxidised iron, were created between 2.3 and 1.7 billion years ago. This allowed the archaeans and their relatives to survive, while oxygen levels in the air and seas rose only very gradually.

Oxygen reached detectable quantities only about two billion years ago, but after that it increased quite rapidly, eventually creating a third atmosphere for the planet.

In this way, a new, aerobic biosphere emerged in competition with the earlier, anaerobic one, and steadily poisoned it. But the organisms that could use oxygen now had access to a new and powerful form of metabolism, one that allowed their cells to co-operate. This led to multi-cellular creatures, the first traces of which date from around a billion years ago. More complex water-dwelling animals, such as jellyfish, worms and molluscs, evolved about 300 million years later. The steady increase in oxygen in the air, and dissolved in ocean water, then set the stage for the Cambrian period, between 542 and 488 million years ago (mya). Within this was the *Cambrian Explosion*, between 530 and 520 mya, when all the basic patterns of modern life forms originated. They did so in water, but there's evidence that some lineages quickly began to occupy the land as well. Afterwards, the variety of life increased as each lineage produced many species, and groups of species called genera and families.

The amount of oxygen in the air had exceeded present-day levels by the Carboniferous period (360–300 mya), allowing the existence of giant insects, such as metre-long dragonflies. These, like all insects, lacked lungs, and depended on the diffusion of gases through spiracles in their bodies and tubes running among their tissues, a design that works poorly in large animals when the air contains little oxygen. The level of oxygen in the air remained high, at about 35 per cent, for much of the 'age of dinosaurs' – the Triassic, Jurassic and Cretaceous periods (251–65 mya) – before declining to a new balance in the modern world.

BIODIVERSITY AND ECOSYSTEMS

Biodiversity is the variety and variation among all kinds of life, in other words the information that has accumulated in living systems over time. It ranges from the genetic coding for proteins, metabolic pathways, cells and individual organisms, to the differences between lineages and species and, ultimately, those among all life forms, relationships and processes in every ecosystem. An *ecosystem* comprises all the organisms living in a particular place and time, all the relationships between them, all the physical features of light, heat, moisture, wind, waves and chemistry that affect them, and the history of the place as well. All ecosystems have a source of energy. In places on the deep ocean floor, the energy sources are volcanic vents that leak heat, methane and sulphur compounds into the icy, lightless water. Archaeans and bacteria use the energy and chemicals contained in the black smoke and red glow of these vents to make more complex molecules with which to live and grow. By doing so, and becoming prey, they feed more complex life forms such as clams, tubeworms and shrimps.

Plant photosynthesis is the dominant energy source for ecosystems on the Earth's surface. By making sugars through this mechanism, higher plants and other photosynthesisers, like algae and phytoplankton, set themselves up as the basic providers for ecosystems. Once photosynthesis had been invented, much of the rest of evolutionary history has been about other organisms trying to steal the plants' chemicals. In the oceans, zooplankton capture and digest phytoplankton, and are eaten in their turn by fish and whales. On land, seed-eating birds eat the starch stores laid down by plants, termites ravage their cellulose fabric, aphids suck their sugary sap, and mammals ranging from rabbits and gazelles to hippos and giraffes graze their leaves or browse their twigs and shoots. The nutrients stored in the bodies of plants and animals are exploited after (or even before) death by a host of fungi and bacteria. Predators hunt down their herbivorous prey. Yet all are sustained by the light of the sun, the water of the soils and seas and the gases of the air, captured and made useful by plants.

ENDINGS

A total of around 100 billion species have existed on Earth, which is roughly the same number as the people who have ever lived, and the stars in our galaxy. But each species usually only survived for a few million years before becoming extinct, with or without living descendants to follow it. There are rare, exceptionally long-lived species, such as the Chinese maidenhair tree or ginkgo, which goes back unchanged to fossils 270 million years old, and has no living relatives. But generally there has been a continual turnover of species, as new ones have arisen in response to new opportunities in the environment, and others have died out because they were unable to cope with change and competition. This is the 'background' extinction rate, but there are also periods during which many extinctions have been clustered together, called *mass extinctions*.

Some mass extinctions were quite devastating, with 70 per cent of all species lost in the event of 350 mya, and 96 per cent of all marine species and 70 per cent of all land species dying in the one of 251 mya. The mass extinction of 200 mya was less severe, taking 20 per cent of all species, and that of 65 mya killed 50 per cent, but these included almost the entire dinosaur fauna, thus setting the scene for the 'age of mammals' (and birds, which are descended from the surviving dinosaurs). The diagram on page 28 gives an impression of the 'tree of life', with each vertical line representing a lineage of organisms containing anywhere from one to a million species. The horizontal bars represent the ferocious culling of lineages in mass extinctions. At least a dozen causes of these past mass extinctions have been proposed, including volcanic events, sea-level changes, the impact of extraterrestrial bodies and sudden or sustained global cooling or warming, all with various mechanisms and all implicated in at least one extinction event. The top-most bar in the diagram represents the sinister influence of people on the tree of life.

THE ANTHROPOCENE

In the modern world, we share our planet with a very large number of other species. No one knows how many, but my own favourite number is 50 million plus or minus 25 million. Sometimes, you'll come across estimates as low as 10 million, or as high as 100 million – or more, since you'll also find reports of discoveries implying that there are 30 million beetle species in tropical forest canopies, or 100 million nematode species in the sea bed alone. But at this level, the numbers don't really matter. The thing to remember is that there are a lot of them, and only about two million have been described anatomically and have scientific names, which in most cases is the limit of what we know about them. The other thing to remember is that we're killing them in vast numbers, either at once, day by day, or by committing them to extinction as their populations and habitats shrivel.

The extreme rate at which species are dying out now will appear in the fossil record of the future as yet another mass extinction. This will be clearly understood, if a successor species exists that's able to understand at all, as having had a completely new cause: humanity. For the evidence millions of years from now will be unambiguous. There'll be an extremely thin layer of rock dividing deeper levels full of diverse fossils from shallower levels with hardly any. The marker layer will contain abundant plastic polymer molecules, radioactive decay products that can only have come from artificial nuclear reactions, and distinctive concentrations of metals. In geological terms it will be called the *Anthropocene*, 'the age of mankind'. Everything afterwards will be known as the post-Anthropocene, just as the three billion years or so before the Cambrian is now known as the pre-Cambrian.

One reason for this bleak outlook is the ecosystem change that's now going on, which is simply depriving wild species of their habitats. But few appreciate the true scale of the mass extinction that is now underway. The Earth's millions of species are not evenly distributed, and about 70 per cent of the terrestrial ones have been found to be concentrated in just 34 *biodiversity hotspots*. Between them, these once occupied about 15.7 per cent of the planet's land area. But 86 per cent of this habitat has already been destroyed, mostly since 1950, and the remnants of the hotspots now occupy only 2.3 per cent of the Earth's land surface. These small and declining patches shelter many species that occur nowhere else: at least

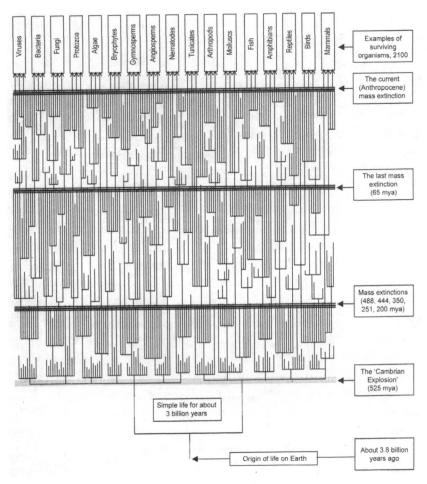

The tree of life on earth

150,000 endemic plant species, almost 12,000 endemic verte-brates, and many millions of invertebrates, mostly unknown to science.

It isn't possible to slash and burn 86 per cent of the habitats of tens of millions of species without at least half of them becoming extinct. Not necessarily at once, but committed to extinction they will be, due to the reduction and fragment-ation of their populations and habitats, and such factors as the deaths of partner species, such as their pollinators and seed-dispersers. If the whole dynamic were stopped today, we would still be looking at millions of species continuing to die out, probably at an accelerating pace as the struggle ends for thousands of ecosystems. This process seems set to peak in the period 2000–2025, when half of the world's species are likely to be lost, at a rate of about a million a year. The continuing growth in demand for farmland, timber and minerals is a major factor in all this, but another is climate change. This is being caused partly by rapid ecosystem change that releases vast clouds of greenhouse gases, but it's mainly due to the burning of fossil fuels, which does the same but even faster. The consequences may well prove to be the defining feature of the Anthropocene.

SOMETHING IN THE AIR

The current atmosphere is 21 per cent oxygen and 78 per cent nitrogen, with trace amounts of other gases, of which the most abundant is water vapour at 0.25 per cent on average (but usually 1–4 per cent near the surface), and carbon dioxide, which I measured in a school science experiment in the early 1970s at 0.03 per cent. This compound has been known to be a greenhouse gas with potential significance for climate change since the 1890s, although scientific studies of the atmosphere and modelling of climate in response to changing conditions and gas compositions really began only in the 1950s. By 1960 the CO_2 concentration in the atmosphere had been measured accurately, at 0.0315 per cent, and an annual rise had been detected. By the late 1960s, scientists were discussing the heating consequences of a build-up of CO_2 in the air, while modelling multi-degree increases in average global temperatures and the melting of polar ice leading to sea-level rise. But over the next few years they discovered the enormous complexity of the climate system, and their models began to include water vapour as a greenhouse gas, as well as the cooling effects of particulate pollution, with contradictory implications for the future.

This lack of clarity meant that the issue of climate change was not specifically considered at the UN Conference on the Human Environment in Stockholm in 1972, which established the United Nations Environment Programme (UNEP). But by the mid-1970s, serious concern was being expressed about the climatic impacts of greenhouse gases, including CO_2, methane and the chlorofluorocarbon gases (CFCs) that had by then been shown to deplete the ozone layer. And by the late 1970s, scientific consensus was well established that global warming caused by human activity posed a serious risk for the twenty-first century. By then, this knowledge had spread widely among journalists and the environmental movement. I was working on tropical rainforest ecology by then, for example, and in 1977 I reported to the World Wildlife Fund (WWF) the dying of mountain forests on Mount Benom in Malaysia, which I attributed to new climatic conditions at high altitude.

The early 1980s saw a political backlash against the environmental movement, especially in the USA, but evidence for global warming and the link with carbon dioxide continued to mount, and this led to an increasing international response. In 1988, CO_2 in the atmosphere reached 0.035 per cent, and the Intergovernmental Panel on Climate Change (IPCC) was formed by UNEP and the World Meteorological Organization to evaluate the scientific evidence and the risks involved. The IPCC published its first assessment report in 1990, and a supplement in 1992 to inform that year's international conference in Rio, which agreed the UN Framework Convention on Climate Change. This provided the political basis for the 1997 Kyoto Protocol on climate change, which entered into force in 2005 and established targets to limit greenhouse gas emissions by some richer countries. Meanwhile, the IPCC produced a series of other assessments, in 1995, 2001 and 2007, all of them based on reviewing published scientific papers. The reports themselves were prepared and reviewed by hundreds of professional scientists – in the case of the 2007 report by nearly 4,000 of them. The reports have consistently increased our understanding of the processes involved in climate change driven by people, and reduced our uncertainty of the likely consequences.

CLIMATE CHANGE

By now, the broad outline, and many of the details, of what's happened and what's likely to happen are supported by a strong consensus among scientists. Key conclusions of the 2007 IPCC report were that global warming is a reality, and that it is almost certainly due to emissions of greenhouse gases such as carbon dioxide, nitrous oxides and methane, from industry, transport, farming and deforestation. Average global surface temperatures have already risen by nearly 1°C over the past 100 years and will continue rising for centuries, even if we stabilised emissions immediately. This warming has had significant impacts on the biosphere (and the role of water within it) and on people. Sea levels have risen only slightly, so far, but enough to threaten low-lying island countries and coastlines with storm surges and sea water intrusion. Glaciers have receded, unusually intense storms and rainfall bouts, and heat waves, have occurred, together with prolonged droughts in Australia, the western USA and north-east Africa.

The rate of warming from now onwards depends on what we do. So the 2007 IPCC report considered scenarios ranging from 'business as usual', with continued rapid, global economic development and unlimited greenhouse gas emissions, to 'global environmental stability', which is more or less the opposite. Thus, relative to today's global average temperature, warming by the end of this century could be as low as 1.1°C in the 'greenest' scenario, and as high as 6.4°C in the 'business as usual' scenario. The social, economic and political implications of making and implementing a choice between these scenarios are so challenging that, in practical terms, the debate on what to do about it is still paralysed by controversy. The dangers of weak leadership become clearer when we look at some of the implications of following the 'business as usual' scenario, with greenhouse gas emissions continuing at the current rate of tens of billions of tonnes annually.

If we take no action, the average global temperature would rise by up to about six degrees during the twenty-first century, and possibly higher and/or faster than this. From our climate models, and from what we know of past climate change and mass extinction events, we can expect biosphere and hydrosphere changes along the following lines for each degree of temperature rise above the pre-Industrial Revolution level. (Bear in mind that there are time-lags in the system, especially since the water of the oceans takes a long time both to heat up and to cool down, so we are already committed to a 1°C increase, and probably also to a 2°C one. There are positive feedbacks involved in such temperature increases too, which can accelerate the warming process.)

At an increase of *one degree*, droughts will plague the western USA and ancient sand dunes will begin to remobilise there, while many dry areas of the world will become drier still. The Atlantic Ocean currents that keep Western Europe's climate relatively mild will start to break down. Most mountain glaciers will melt, and Arctic and Antarctic ice cap melting will be well underway. Most of South-East Asia will become dominated by fire-maintained grassland, and remaining rainforests elsewhere will start to die. Most coral reefs will bleach, die and become algae-dominated ecosystems instead. Sea levels will continue to rise, and hurricanes and other large storms will become more intense.

At *two degrees'* increase, dissolved carbon dioxide would cause oceanic acidity to increase to the point of weakening the surviving coral reefs and making it hard for marine organisms to build shells, and therefore it will start to close down phytoplankton photosynthesis. More than half the summers in Europe would be hotter than the heat-wave summer of 2003, which killed 30,000 people. Heat-wave conditions would become established as normal in southern Europe, and forest fires and water shortages would become frequent throughout the Mediterranean basin. Sea-level rise would accelerate, threatening coastal cities. Ecosystems everywhere would change their distributions, or die where they stand, and vast numbers of species would become extinct, along with many human cultures. Storms, droughts and floods would all become more severe. Food shortages and famine would become frequent and widespread, especially in Africa.

At *three degrees'* increase, the Amazon rainforest would dry and burn completely, and be replaced by desert. The further drying and heating of drought-prone areas such as southern Africa, north-western and central America, and most of Australia, would render them uninhabitable and largely covered by shifting sand dunes. Remaining glacier ice would melt completely, almost halting the flow of rivers that rely on glacial melting for much of their water, such as the Indus, Ganges/Brahmaputra, Mekong, Yangtze and Yellow, which together sustain half the world's current population. Storms and storm-surges would begin to engulf low-lying cities and countries. The Sahara Desert would enter southern Europe. Most of those species not killed by land-use changes in the late twentieth and early twenty-first centuries would die out, and 'the age of loneliness' would have begun.

At an increase of *four degrees*, the rate and extent of sea-level rise is currently unpredictable, being between 1 and 25 metres at various times, depending mainly on the behaviour of the Antarctic ice sheets. While large areas in the global north and south become deserts, the monsoon zones of East Africa and parts of India would be wetter and hotter, more disease-ridden and with few crops adapted to the new conditions. Otherwise, the world's weather would be extremely agitated, with storms of incredible ferocity striking increasingly wide areas, and deserts spreading across Europe as far as southern Russia. Huge areas of Europe would be abandoned by refugees heading for the storm-battered regions of the Baltic, Scandinavia and Britain.

At *five degrees* hotter, the Earth would be largely unrecognisable as well as mostly uninhabitable: the shapes of continents drastically altered, all the ice and rainforests long gone, fires raging across the far northern forests of Canada and Siberia, the interior of continents on average ten or more degrees hotter than now. Surviving people, if any, would be fighting over the remaining habitable lands.

These scenarios are gleaned from Mark Lynas's nightmare-inducing book *Six Degrees*, which is based on work mostly published since 2000 in refereed scientific journals and by university presses (many of the same sources used by the 2007 IPCC assessment). The 6°C world is largely indescribable other than in Hadean terms. The last time the Earth was as hot as that was during the mass extinction of 251 million years ago, when almost all life died out. The problem is, though, that feedback mechanisms are known to exist that are likely, in a 5°C world, to make a 6°C world inevitable, in this case the release of huge amounts of methane from gas hydrate deposits in the continental shelves. Similarly, the 4°C world would create a 5°C one through the release of methane and carbon from melted, decaying and burning Arctic permafrost peats. Then again, the 3°C world would ensure a 4°C one through the reduced reflectivity of a dark, ice-free Arctic ocean, the breakdown of carbon cycles and the smoke and ash of fires. And at 2°C, the ending of the capacity of the oceans to absorb further carbon dioxide would ensure the transition to 3°C.

To be reasonably sure of keeping below a 2°C temperature increase and avoiding the worst effects of runaway warming, we would need to stabilise the carbon dioxide concentration in the air, more or less at once, at about the level that it currently is – 0.04 per cent. We would also have to keep a tight rein on emissions of other greenhouse gases, like methane, and bear in mind that water vapour itself is a potent greenhouse gas, which we are releasing by pumping water out of deep aquifers where it has long been isolated from the atmosphere.

This book is about water, not climate change, but it is clear that crises of water scarcity, distribution and quality must be related to ecosystem damage and climate change, and that biodiversity loss is linked to both, so all these processes feed off each other, and amplify one another. I assume that, as the scale of the climate change threat to our existence becomes widely known, an irresistible demand for leadership will arise, to allow fair, radical and effective measures to be taken. I think that everyone who knows what's going on is waiting for such leadership, and will support and participate willingly in the necessary changes, no matter how demanding, as long as they're fair. While this momentum builds up, though, it's important to be aware that everything's connected, and that greenhouse gas emissions are only one part of the big picture. Ecosystem destruction and mass species extinction are underway as distinct phenomena as well as causes and consequences of climate change.

GAIA CONNECTS

It's clear that life has evolved within the atmosphere, oceans and rocks of the Earth, with each affecting the other. Despite all the radical changes in the air since life arose, adjustments and feedback arrangements have contrived to keep the surface of the Earth within a narrow range of temperatures over hundreds of millions of years. This range was just the one needed to keep water liquid, and therefore to keep life alive. If life on Earth ended tomorrow, oxygen would vanish from the air in a few million years, sucked out by inorganic chemical reactions, while carbon dioxide would be dissolved in water and be absorbed into rock in a few thousand. These gases have to be continually maintained by biological activity and geological processes, which seem to co-operate in maintaining reasonably constant levels over vast periods of time.

A similar constancy is seen in the composition of the oceans, which has remained the same during geological time at a salinity of 3.5 per cent, with the proportions of various salts held in exquisite balance. This precise composition should have changed drastically with the erosion of salts from the land, but it hasn't. Looking at this overall, long-term pattern of dynamic balance in the Earth's systems, in the early 1970s James Lovelock introduced the idea of *Gaia*, after the ancient Greek goddess who personified the Earth. This he described as a complex entity involving the biosphere, atmosphere, oceans and soil, all of them parts of a feedback system that maintains conditions exactly favourable to life. His collaborator, Lynne Margulis, described Gaia as the single huge ecosystem at the Earth's surface, which is made up of all the connected ecosystems there, and which behaves in some ways as a kind of physiological system.

GAIA CORRECTS

To an ecologist equipped with hindsight and a Gaian education, all this seems quite reasonable. There's no particular reason to see Gaia as either a real goddess or a real super-organism, but in practical terms you wouldn't be far off track if you did. In any case, conscious or not, the capacity of Gaia to correct imbalances in the biosphere is clearly vast. The planetary over-heating caused by our emissions of greenhouse gases is certainly on a scale likely to provoke such a correction. Quite what it might involve is not clear, but on past performance we'd expect the causative agent of instability – our own excessive numbers and impacts – to be especially affected.

The blunt responses of Gaia cannot be precisely targeted, and as the correction occurs we'd expect the present-day biosphere to be utterly transformed. But, alone among all the species that have ever existed, we have the capacity to anticipate this. We could, in theory, act to transform our own relationship with the biosphere, instead of letting it change under us. The global water crisis described in the rest of this book amounts to a warning that transformation is inevitable, one way or the other. That is, transformation either of the biosphere, or of our attitudes, behaviours and economies.

First, though, we should have a look at where the attitudes that influence our behaviours might have come from. This will help us assess what flexibility we might be able to call on from our own evolutionary heritage, and what opportunities we might have to solve our ecological problems. As we'll see, there are natural capacities buried within us all that we can mobilise, and that can help us greatly. This is my main reason to be optimistic about the future, and the main reason why I wrote this book.

3. THE EXPERIENCE OF WATER

SENSUAL WATER

A diver rolling backwards into the sea, or a snorkeller swimming out over the reef edge, is at once immersed in a new world that seems designed to grasp and hold the attention of a human. There is the sense of flying, a feeling that our remote ancestors must have wondered about ever since they cooked their first bird. Then, in the sea there is a blueness, part of the light spectrum to which we are very sensitive, and also twilight, which makes us alert. These responses may reflect the many generations our ancestors spent trying to spot predators in semi-darkness, whether hyenas in the savannah or tiger sharks in the shallow sea. Everything looks bigger in the sea, too, making it that much more interesting. When we enter the ocean, we plunge into a shadowy world where we are surrounded by magnified alien life forms, many of them looking at us.

The abundance of marine life is part of the fascination. Our ancestors, and later generations of beachcombers, coastal foragers and spear-fishers, would have seen a thriving sea bed and dense shoals of fish as a welcoming resource, somewhere to stay awhile and use thoroughly before moving on. They would have swiftly evaluated the density of life, spotting telltale indicators of harvestable creatures. They would have looked for the siphons of buried clams, and for sea cucumbers, urchins, snails and chitons. They would have assessed the abundance of the particular fish that they knew to be slow enough to be caught, spineless enough to be handled, and free of poisons enough to be eaten. The sheer quantity of edible things would have been the first draw, and an appreciation of the diversity of life may have come later.

Away from the coastal waters, our ancestors on land would have been especially interested in places with lots of living things of many different kinds, since these would have indicated water, fertility, prey and plant foods in an arid environment. Thus we have inherited a 'biophilia' instinct – a love of life which makes us put plants in our apartments, pay extra for houses near botanical gardens, and go 'Oh, wow!' when we see all the different fish beneath the ocean's surface. This instinct helps create our preferences for different places. Another influence is that we love good visibility, places where we can see a long way in safety. This is surely another relic of our past, and a good reason for divers and snorkellers to prefer 30-metre to 3-metre visibility.

Then there's the excitement. There are plenty of things that are, or that look, dangerous in the sea. Most of the danger is illusion, but some is real. Overcoming danger before witnesses has rewarded many people for many generations, so its fascination for us is not surprising. It also builds comradeship. We are social animals, and we seldom use the seas alone. Indeed, the first rule of diving is never to do it alone. During dive training, you have the instructor to guide you through the dangers. Later on, you have a divemaster or guide to refresh the sense of being led and looked after, and initiated into new mysteries. Co-operation has other rewards, too, since diving is a great way to make friends with total strangers. Anyone with a dive bag is fair game. You just walk up and say something about diving, and very soon you're kitting up with them and their friends at a dive site, with a shared adventure ahead and socialising afterwards.

Our long history in the water and on dry land has prepared us well to swim and frolic along the coasts of our world. Our eyes, minds and social priorities are all pre-adapted to appreciate the marine experience. We are interested in the sea and its life forms, we find it thrilling or pleasant to be immersed in water, we enjoy the hot sun warming us after swimming in cool water, we are stimulated by new social interests, and we slip into a relaxed and positive mood in response to marine sounds, smells, foods and drinks. So it's not surprising that wanting to be by the sea is the backbone of tourism, a sector that generates more than a tenth of the world's economy, employs 200 million people, and transports 700 million international travellers per year, a number that could double by 2020. Tourism is one of the top five exports for 80 per cent of all countries, and the main source of foreign currency for half of them. Simply put, tourism is one of the major activities of people, and at its heart is water.

SACRED WATER

People have appreciated water since ancient times as something more than just for use, either for recreation or as a daily necessity of life for themselves, for their livestock and prey, and for their cultivated plants. The Chinese 'way', or *Tao*, is represented by the winding course of a river, with water acting as a unifier of the 'female principle' (*Yin*) and the 'male principle' (*Yang*). The sacred use of water is embedded in the purifying ablutions of Hinduism, Islam, Judaism and Shinto, the sprinkling of Holy Water, and the baptismal ceremonies of Christian and non-Christian peoples. We've given water a spiritual potency for as long as we've had abstract thought and the language to express it. We've often extended this to the inhabitants of water too, such as fish in general and sharks and eels in particular. In Jungian dream amplification, water is considered a key symbol of soul or life-spirit.

It's not surprising, therefore, that watery places have often had a sacredness about them. The gentle splashing of water into a shady pool seems to have an irresistible drawing power, especially if the pool is set amidst tall and shapely rocks, and is approached by a winding path that allows the visitor to contemplate mysteries. In every culture, places like this accumulate offerings, prayers and such things as good-luck ribbons tied to bushes. The archetype is the Castalian Spring in the ravine between the Phaedriades, a pair of 700-metre cliffs on the lower southern slope of Mount Parnassus, at Delphi in Greece. This spring was once believed to have been guarded by the fearsome serpent Python, who was later killed by the God Apollo. Here, for many centuries, all those who came to consult the Oracle of Apollo stopped to wash their hair.

Look at Paul Gauguin's painting *Nave Nave Moe* or *Sacred Spring*, which is set in Polynesian Tahiti. Here there is everything a sacred water place should have: peace, shade, rocks, trees, fruits and, of course, water flowing from a mysterious source to meet the needs of the people. If such a spring lay in the world of the Zuñi, in New Mexico, it might have had a wall around it to keep out the cattle; if in the ancient Roman Empire, like the one at Bath in England, it might have been encased in stone buildings and temples of the Goddess Minerva; or if in the Catholic world, like the one at Lourdes in France, it might have become a mighty place of healing and of pilgrimage, thronged by the desperate and the devout. For sacred water can also be healing water.

HEALING WATER

Our ancestors must have cut themselves often, wading around in tropical waters, but sea water would have cleansed their wounds. Maybe then, or perhaps much later, the association of bathing with healing was established. When a wounded warrior returned from battle, sweaty, dirty and crusted with blood, he would have been washed before his wounds were bound. After that, his fate would have been up to luck and the robust immune systems of his day, challenged since childhood by cuts and scrapes. If he recovered, maybe the water would be credited, especially if it had come from a special place and been kept for this purpose, or blessed by a community priest or holy man.

Since the time of Ancient Egypt, water has been seen as part physical and part spiritual, or else as somehow energetic. Even today, there are many who believe that water's mysterious dimension allows it to carry information and medicinal potency. Homeopathy, for example, is based on the idea that water can 'remember' the pattern of a curative substance that was once dissolved in it. Cranial osteopathy regards the body as being connected mechanically, biochemically and energetically through water bound to collagen in the connective tissues that permeate and structure the whole body. Osteopathic treatment is said to flip this water from a structured gel state to a fluid sol state, where curative adjustments can be made, using a mysterious potency which practitioners call the *Breath of Life*. Classical acupuncture calls this same energy *Qi*, which is also used in many other ways through diverse disciplines, including *Taiqi*, *Reiki* and *Qigong*.

All these uses of water and its unknown energies seem to be real and effective at some level, but the nature of the energy involved and the precise mechanism of its action are poorly understood, and are often even unrecognised by scientists. Our problem is a lack both of instruments, other than living organisms themselves, to detect and measure *Qi*, and a theory to relate it to everything else we know. Meanwhile, water flows on, laden with potentials and mysteries, if only we knew what it was and how it worked. We know that there is some kind of extra feature of water, since *dowsing* for it works, but no one understands dowsing either. Whether this extra dimension is a function of the shifting balance of electrostatic charge among different combinations of water molecules under different conditions, or is pure *Qi*, it is clear that there is indeed something extraordinary about water, awaiting only genius to explain.

But what is this human species that has such strange ideas? Perhaps we can shed some light on our experience of water, and our relationship with it, by looking further at our evolutionary history. For one thing's certain: our entire life as a species, and the lives of all our ancestors back to the furthest reaches of evolutionary time, have been spent in intimate contact with water: needing it, seeking it, using it, reacting to it. And this must have left its mark.

HUMAN ORIGINS

We belong to the hominid family of mammals, together with six close relatives. There are two species of orang-utans, in Borneo and Sumatra, which we might call second cousins, and two gorillas, in eastern and western Africa, which we could think of as first cousins. Closest to us, though, almost brothers really, are the two chimpanzees, the bonobo, from south of the Congo River, and the common chimp, which lives across Africa north of the Congo. The family home is mostly tropical rainforest, but common chimps have also spread into drier forests, and we ourselves have occupied the world. All seven of us can be traced back, using fossils and molecular 'clocks' based on the rate of change in our DNA and proteins, to a common ape ancestor, the first creature of the monkey kind that wasn't actually a monkey, which probably lived 22–25 mya. After that, the lineage that eventually produced the orang-utans spun off 10–12 mya, the lineage of the gorillas did so 6–8 mya, and the lineage of the chimpanzees split away from our own 4–6 mya.

This last separation almost certainly happened in Africa, but it's important to bear in mind that Africa was a very different place then, with climates and ecosystems unrecognisable and changeable relative to conditions in the modern world. Moreover, we have no real idea what the last common ancestor of the chimpanzees and people looked like, how they behaved, or why the two lineages went their separate ways; all this is hotly debated by specialists. Nevertheless, we have built up a reasonably detailed picture of the evolution of our bones and teeth at least, since we parted company with the chimpanzee lineage. In these studies, we have focused mainly on evidence from fossils that sheds light on what we see as the key differences between us and them, i.e. the great size of our brains, which began to expand about two million years ago, and our design for walking bipedally, on our hind legs.

How, why and when we began walking on two legs remains very uncertain, and the field remains open for theorising. In May 2007, for instance, scientists speculated that early pre-humans began doing this in trees, as a way to reach fruits overhead, and were therefore able to walk on the ground when climate change took away the forests. This is a variation on the usual vision of our ancestors becoming upright in the African bush-savannah, having first gone through a phase of knuckle-walking like chimpanzees, and then striding into their destiny as uniquely bipedal mammals. While there is evidence that pre-humans occupied African savannahs, there are puzzling aspects to this story that must be explored in a book about water, and particularly in a chapter on the human experience of water. This is because the puzzles all suggest that our lineage may have adapted to life in water, as well as to life on dry land.

AQUATIC APES

Most of the evidence for a semi-aquatic aspect of human evolution comes from our anatomy and physiology, which in many ways are so unlike anything to be found among the rest of our great-ape family. In contrast to those of our closest living relatives, the human form reminds us of how marine mammals are designed – animals like dolphins, whales, manatees and seals – and there are other similarities with wallowing mammals like the hippopotamus and babirusa, and even with penguins. The obvious conclusion, that our lineage must once have been semi-aquatic, was reached in the 1930s independently by Max Westenhöfer in Germany and Alister Hardy in England. Hardy was a professor at Oxford University from 1946 to 1961, as was J.R.R. Tolkien between 1925 and 1959. I mention this because they may well have influenced each other in developing the character of Gollum in Tolkien's *The Lord of the Rings*, who is described as a semi-aquatic, fish-eating, ancestral form of the more human-like hobbits. In any case, Westenhöfer's and Hardy's basic idea, now called the aquatic ape theory, has been further developed since the 1970s by Elaine Morgan. She has done this in the face of opposition from more orthodox scientists, who believe that we became human in a hot, dry environment, where we developed our bipedal walking, language and tool-using skills as adaptations to a terrestrial life as hunters and gatherers. The aquatic ape theory, though, marshals the following lines of reasoning.

1. **Hair.** What hairs we have on our bodies are generally fine and leave most of the skin visible and exposed. This arrangement is typical of marine and wallowing mammals, but very different to the other apes or any other terrestrial mammals that live exposed to hot sunlight and dry air. Our hair is also aligned in ways that imply a certain streamlining, consistent with an adaptation to swimming. The continuously growing hair on our heads, however, might be interpreted as an anchor for babies swimming or floating around us, as well as a shield against hot sun when we are otherwise submerged for hours at a time. The distinctively textured hairs of armpit and crotch, then, are seen as being tufts for dispersing scents and evaporated hormones (pheromones) from skin glands, developed when we later returned to dry land where scent signals would again be useful in communication.

2. **Fat.** Our nakedness raises the question of insulation, since most mammals keep warm or cool by trapping a layer of air beneath their fur. Insulation is particularly important in water, which conducts heat far better than air does, and waterlogged fur is no use in keeping warm. Instead, water-living mammals that are too big to trap a single sheath of air against their skin typically rely on a layer of fat beneath the skin. Unlike any other primate species, which tend to store fat in the abdominal cavity, humans too have such a layer of under-skin fat. This implies that keeping warm during long immersions in water was an important challenge for our ancestors, which was solved in the traditional way for marine and wallowing mammals. This fat layer also increases streamlining, which would aid

swimming and diving. Fat is also important in buoyancy, since it helps people float on the surface but is also, during repeated breath-hold dives, compressed by water pressure to reduce buoyancy, making it easier to stay down. Finally, human babies are born very much fatter than ape or monkey babies. One can imagine them bobbing plumply to the surface on being born, where they are guided to the mother's breasts, which are also fatty and float conveniently on the surface.

3. **Sweat.** If we use under-skin fat to keep warm in water, we use sweat to keep cool in air, secreting this salty fluid all over the skin where it evaporates and takes heat away from the body. Again, among the primates and most other mammals this is a uniquely human thing to do. It is also extremely costly in terms of the two resources that are most scarce in hot, dry environments: water and salt. Aquatic ape theorists would argue that these same resources are, by contrast, abundant in sea water.

4. **Kidneys.** Our kidneys are unique among primates (and rare among purely terrestrial mammals) in having an inner layer (the medulla) that comprises multiple lobes and pyramids. This design increases the surface area of the tissues that function to clear salts and nitrogenous wastes from the blood into the urine. It is an arrangement universal in marine mammals, and common in non-marine aquatic and terrestrial mammals that are thought to have had marine or coastal ancestors.

5. **Walking on two feet.** We are the only mammals that normally stand and move only on their hind legs, an arrangement that has demanded many design changes in our bodies. It is also a slow and unstable posture, and therefore dangerous because of the risks of predation and falling over. People cannot out-run savannah predators like lions and hyenas, and tripping or slipping one's way to a broken wrist or skull is a common fate for humans. The aquatic ape theory makes light work of the problem, though, for a vertical posture makes perfect sense when wading in water. This is stable, keeps the face well above

the surface, and blends easily into vertical floating or treading water. One can imagine watching from the shore a group of our ancestors who have drifted out to sea, only their heads and hair-anchored babies visible, some ducking and diving for shellfish. After a while, they lie over in the water, keeping their spines in line with their legs as they swim back to shallower water, kicking their webbed feet, before regaining their footing and wading further along the coast.

6. **Swimming.** After an initial reflex tendency to hold on to something, human babies have a natural aptitude for floating, paddling about and swimming in water, and for holding their breath when their faces are immersed. It seems that we can swim before we can even crawl, let alone walk, and this is an unlikely adaptation to an early life on dry land. Birthing in warm water is now accepted to be less traumatic for both baby and mother than conventional arrangements involving dry beds and bright lights. Even if not born in water or released there as a newborn, most people can easily learn to swim and few exhibit the overwhelming fear of water that is so common among our great-ape relatives.

7. **Diving.** People can breath-hold dive to a depth of tens of metres, similar to the foraging depths of dolphins, penguins, seals and walruses, and can stay down for several minutes. These abilities improve with training and experience, so it is not hard to imagine that our ancestors in coastal waters would have been able to exploit a wide range of sea-bed resources.

8. **Speech.** Being able to speak is special to humans, and although great apes can learn to communicate with each other and with people using sign language they can barely learn to use spoken words despite possessing a naturally rich vocal range. The key is the conscious control of breathing, which is essential to diving animals and equally essential to making complex and varied vocal sounds. Aquatic ape theorists argue that life in and near the sea would have given our ancestors the basic capacity to

develop speech as a primary form of communication. This may not have become fully developed until much later, with the expansion in our lineage's brain size, but some visualise aquatic apes using speech to co-ordinate group activities, such as herding shoals of fish into tidal pools where they could more easily be caught by hand. Dolphins, too, demonstrate the importance of vocal communication and co-ordinated action in concentrating prey in a three-dimensional environment, one in which it is all too easy for fast-swimming fish to slip away unharmed.

9. **Tears, love-making and noses.** There are other features that seem to align us with marine mammals rather than with our close relatives or other terrestrial species. Tears, for example, since, apart from elephants, humans are the only terrestrial mammals that weep when emotionally disturbed, but this is common among marine mammals such as seals and the now-extinct Steller's sea cow. Or a good sense of balance, which we have along with marine mammals such as seals, sea lions and dolphins; like all of them, we can learn to balance balls on our noses, which is not something that any of our close relatives can do. Making love face to face is our most common preference, as it is among whales, dolphins and manatees (and was among Steller's sea cows), which, like us, are designed for it. But this is very rare among terrestrial mammals: bonobos and orang-utans are exceptions, and this, at least under some circumstances, is to do with their distinctive social systems. Even our noses seem well designed to exclude water if we dive head first into or under water. If we jump in feet first we tend to block our noses with our fingers (which, by the way, are webbed in some individuals, though not as often, at about 7 per cent, as our toes).

Thus it seems quite likely that the aquatic ape theorists are on to something, and that our ancestors did spend some hundreds of thousands of years in a coastal environment, living a semi-aquatic lifestyle. There are even grounds to identify a place and time for this phase of evolution. This is based on the reasoning that a population of ape-like beings must have been isolated from the broader lineage in an island or archipelago, most likely by rising sea levels between 5 and 10 million years ago. This is towards the end of the Miocene, an epoch that ran from 26 to 5 mya and was a time when ape lineages were very widespread and diverse. The late Miocene coincides with the split between the lineage that led to humans and the one that led to chimpanzees. One leading candidate for aquatic isolation at about the right time is the Danakil mountain range, at the southern end of what is now the Red Sea. This was an island from 7 to 5 mya, because of flooding in the African Rift Valley and the Afar Depression.

The idea is that the apes, isolated on an island some 540 km long by 75 km wide, would have diverged from their common ancestor with the chimpanzees by adopting, and adapting to, a coastal and semi-aquatic marine lifestyle. When sea levels fell again, these beings would have been able to explore southwards, keeping close to water, along the moist Rift Valley all the way into East Africa and beyond. By that time, our ancestors would have been adept at walking on two feet, probably capable of using very simple stone tools after millennia of breaking open shellfish, and may have been able to talk well enough to co-ordinate hunting activities, or at least would have had sufficient vocal and breath control to make this possible quite soon. They would also have had, unlike the chimpanzee lineage, an abundance of enzymes for breaking down animal proteins, which would previously have been driven by the high proportion of aquatic animals in their diet, but which also pre-adapted them for hunting. Thus, despite the disadvantages of sweating, nakedness and fatty insulation, they were able to survive, spread and diversify until, at about two million years ago, one lineage among them developed the large brain size that precipitated the extermination of all rivals, and the eventual conquest of the world.

Their later spread was probably accomplished largely along the coasts of the Old World. Here they could have waded, swum and foraged freely all the way from Africa to the Far East, maintaining the usefulness of their aquatic adaptations all the way, but also advancing up river valleys as they went, and reaching Java by about a million years ago. Their descendants afterwards, with waves of mental, cultural, technological and anatomical innovations, local adaptations, invasions, replacements and extensions, spread out through the whole of Africa, Europe, East Asia, South-East Asia and, eventually, about 100,000 years ago, reached Australasia, and later the Americas. The details may be in dispute, and some of the story changes with the finding of each human-like fossil, but it does seem likely that the key features that make us human had their origin in ancestral adaptations to the coastal seas. These adaptations empowered our lineage to thrive near seas and rivers, and make do, at least better than our competitors, even far inland.

WATER AND A DIVIDED HERITAGE

Most of the puzzling features of people's bodies, like hair, fat, sweat and kidneys, are unlikely to leave a trace in the fossil record, though as we've seen these 'scars of evolution' can be compared with those on other living creatures, and inferences drawn. Behaviour, attitudes and social relationships also don't fossilise, yet their evolution is of particular interest if we want to find out whether our nature might help us solve our ecological problems. We might be able to shed some light on this, if we consider what behavioural and mental marks might be expected from a semi-aquatic dimension to our evolution, as compared with a terrestrial one.

An aquatic lifestyle should encourage co-operation, such as in herding fish, and spotting and deterring predators. It should also promote egalitarian gender relations, since it's hard to control or oppress another individual in water. It might also be expected to encourage three-dimensional movement and lateral thinking, as well as floating and relaxed enjoyment. An attitude of going with the flow would be encouraged by learning to use currents rather than fighting against them, and accepting that fierce storms may occur at any time, inexplicably and unpreventably. Such a lifestyle would tend to discourage ambitions of territorial conquest, since the environment is fluid and ungraspable, immense in scale, and generously productive of food items for roaming, harvesting beings.

A terrestrial lifestyle, though, would be expected to leave different scars in our minds. Here co-operation and solidarity would still be in demand, but a hard, horizontal, linear world offers much more opportunity for physical control of space, of resources, and of each other. So, on land, we'd expect hierarchical dominance systems to develop, with adult males competing with one another for status and the control of females, food and water. Territorial aggression between groups would also be anticipated, led by males and aimed at the elimination of outsider males and the control of space and resources. From this point of view, relatively 'soft' aquatic apes, with their gentle and egalitarian ways, would inevitably have become 'harder' on dry land, and this set of influences too would be expected to have left an imprint on our minds.

WATER AND THE DIVIDED MIND

If we have had both a semi-aquatic and a fully terrestrial evolutionary experience, we'd expect people to show signs of both, and in particular to be able to think and behave in ways appropriate to both. And this is in fact the case, since humans seem equally adept at living in either of two alternative and contrasting models of society: one militaristic, controlling, male-dominated and hierarchical; the other peaceful, accepting and egalitarian. It seems we can do *either*, depending on circumstances. So our minds must be inclined, and have the capacity to think, in two contradictory ways: a hard way, and a soft way. Their different implications are expressed sometimes this way, sometimes that, in response to social context, lessons learned in upbringing, and the observations, reasoning and self-discipline of which people are capable during their long lives.

Let's say that the 'hard' side of our mind is associated with command and control, status, hierarchy, dominance and the expectation of submission by our underlings. This way of thinking shows itself in the ideals and social arrangements of military and imperial societies. It feeds into the ways that such societies order themselves into castes and ranks, how they relate to others through war, threat, tribute and terms of trade, and how they train their youngsters, stressing respect for elders and superiors, their place in society and defence of the status quo. It also relates to how they manage their environments, with the wonders of nature valued only to the extent that they are resources for use by the élite, and the objects of the physical world and the systems of the living world both being seen as controllable through force and engineering. Let's call this approach 'Confucian', as it signals the dominant conservative ideology of Imperial China. But it might just as well be called imperial, top-down, mechanistic, reductionist, or just 'hard' thinking. It is the thinking of a terrestrial ape.

Now let's say that the 'soft' side of our mind is associated with a much more organic approach to the world, one that accepts its complexity and subtlety, that values diversity and the individual lives of all its citizens more-or-less equally, even if they don't happen to be human. In this approach, the invasion, oppression and exploitation of groups of people is uncomfortable, and maintaining such arrangements by training youngsters to accept them is unthinkable. Likewise, the short-term and destructive exploitation of nature is unattractive, and opposed on the grounds that the future is just as important as the present, and there is no recognition that claims on resources by élites outweigh the claims of others, even if they happen not to have been born yet. Physical objects may be gently improved through artistry, but not crushed and fundamentally altered. Natural systems may be collaborated with, but not diverted, felled, dynamited or polluted into a different state of being. Let's call this approach 'Taoist', as that's the philosophy that has long competed with the Confucian orthodoxy of Imperial China. Again, however, it might just as well be called liberal, bottom-up, organic, holistic, or just 'soft' thinking; the thinking of an aquatic ape.

It's hard to imagine philosophical traditions more at odds with one another than Confucianism and Taoism. Confucianism is rational, active and dominating, while Taoism emphasises all that is intuitive, mystical and yielding. Both seek social harmony, and harmony of mind, but by very different routes, and only Taoism explicitly seeks harmony with nature as well. Nevertheless, they and all their implications and consequences both come from the mind of one species. They represent two competing, yet subtly complementary ways of looking at the world and acting within it. As the heirs to both, we humans have at various times used both 'Confucian' and 'Taoist' ways to make a living. The next few chapters describe how these approaches have led to very different strategies and outcomes in our efforts to manage water and water-bearing ecosystems. We'll look at our use of water and living things in the oceans and in swamps, lakes, rivers and the ground. In each case, we can bear in mind the distinction between 'Confucian' and 'Taoist' as labels for different approaches to water and ecosystem management.

4. OCEAN WATER

A wandering albatross soars over the rolling crested waters of the southern ocean, very far from land. She's been airborne for weeks on her great wings, over three metres from tip to tip, aside from sudden plunges to snatch squid or fish from the waves. At a cruising speed of 55 km/hour, she's pushed along by constant wind at about five metres above the sea, and she misses few opportunities to feed. She passed a ship twenty minutes ago, but she's already forgotten it; she's seen many and cares nothing. Suddenly she catches a gleam in the water, a flash of silver scales in the afternoon sun, and she veers and dives as fast as thought. In an instant and a brief splash the target is snapped up and she rises again into the air. But something's wrong: her head is yanked back and she tumbles after it in a clumsy tangle of wing-beats, the object in her beak hard and sharp and inanimate. She plunges into a wave, her throat filled with sea water, and finds herself being dragged along just below the surface. Far away, the long-liner is reeling in her twenty kilometres of monofilament nylon with its freight of steel bait-hooks, tuna and sea birds. The albatross lives for a few long minutes more.

THE CYCLE OF LIFE

About 97 per cent of the 1.4 billion cubic kilometres (km^3) of water on Earth is sea water, and the oceans cover 71 per cent of the planet's surface. There are three great ocean basins, the Indian, Atlantic and Pacific, the last two fusing in the far northern Arctic Sea, and all three combining in the southern ocean to encircle Antarctica. These vast bodies of water influence every aspect of life on land. Every day, megatonnes of water evaporate from their surfaces, and every 3,100 years or so, only as long ago as the destruction of Troy, a volume of water equivalent to all the oceans passes from liquid water into the air, and back again. At the temperatures prevailing on the surface of the Earth, this is extraordinary behaviour for any substance, but it's what makes the land habitable at all.

Last Tuesday, there were about 13,000 km^3, or 13 trillion tonnes, of water in the Earth's atmosphere, an amount of vapour more than six times greater than the quantity of liquid in all the rivers in the world. The same will be more-or-less true next Wednesday, and every day, except that the world is becoming warmer each day so another hundred thousand or so cubic metres evaporate. These vast amounts of water vapour are carried in the warm, moist air that rises above the sea, but as they rise they cool, and the vapour condenses into droplets of liquid water, forming clouds. The droplets may freeze if the cloud rises high enough, but as solid or liquid they drift with the winds until the physical conditions of tempera-ture and pressure are just right for the droplets to coalesce. And then they fall, as rain or snow, about 500,000 km^3 of them every year, either down into the sea, or onto land.

Once on land, about 100 km^3 daily runs off as rivers, and 160 km^3 evaporates again, either directly from rocks and soils, or having been sucked up and used by plants before being released from their leaves. Some is delayed for a while (sometimes a long while) as terrestrial ice, but all of it eventually falls again as rain, on land or sea, or else rejoins the sea via a coastal glacier. There is a stupendous cycle going on, with the sun's heat drawing water into the sky, and the cool atmosphere distilling it back to the surface. Thus everything keeps moving, and living.

SEA LIFE

From our terrestrial viewpoint, we tend to think of biological diversity in terms of trees and ferns, birds, insects and mammals. But unless you venture beneath the sea you will never encounter half of the fundamental kinds of living creature that exist on our planet. A quick sample of the life clinging to a rocky wall ten metres below the surface of a tropical sea will reveal many species of corals, sponges, tentacled anemones, sea squirts and algae.

In daylight, a spectacular array of swimming creatures glides and darts about. Blue triggerfish dot the sea up to about five metres from the wall, plucking plankton from the water. Sometimes, for no apparent reason, they sense something that makes them shimmer back all at once to hide by the rock face, before cautiously emerging again. An imperial angelfish man-oeuvres its blue-and-gold striped body until, almost upside-down, it can get its mouth to some tiny item of prey under a ledge. Further along the wall, a great barracuda hangs motionless, surveying the prospects among the gaudy tiddlers by the wall. Then it moves without warning, and, instant-aneously, all that remains of a silver jack is a cloud of scales. Overhead, a green turtle is silhouetted against the silvery underside of the surface, as it takes a breath and then vanishes over the crest of the wall. A torrent of small, bright lunar fusiliers cascades down the rock face, as a white-tipped reef shark crosses their path.

All this changes at night, as different creatures take over. Everywhere is the shine of shrimp eyes in the torchlight, like the eyes of so many rainforest spiders. Basket-stars have emerged from their shelters and clambered to the edges of coral and rock, unfurling their writhing tentacles into complex fans that are held concave to the passing current. Crabs are active, some with sponges and anemones glued to their carapaces. There are mantis shrimps, slipper lobsters, delicately branching gorgonians, feather-stars, spiny lobsters and parrotfish sleeping in bags of mucus in little caves. Lionfish hang in the water, surrounded by diaphanous fins and poisonous spines. There are brittle-stars creeping, sea-stars feeding, and urchins waving their pencil-thick spines. The flat disks of *Fungia* coral, dead-looking by day, are now covered in erect tentacles and crawling around. Any agitation in the black water stimulates plankton to glow like constellations, and the turds of alarmed fish are luminous green as they dart away.

The sea provides a much more three-dimensional environment than the land, with a far greater volume of space available for its inhabitants. This space is extremely varied in conditions of light, temperature, salinity, nutrient concentration and pressure. The physical structure and lifestyle of every marine organism reflects its place in the ocean. In mid-water there may be a floating pink siphonophore, looking like a translucent eyeball with a blue iris. It pulsates up and down as it drifts, controlling its buoyancy by secreting carbon monoxide into its float. Behind it trail some ten metres of fine, contractile, stinging tentacles, a deadly trap for prey. This gelatinous design is allowed because sea water supports living tissue, so floating or weak-swimming marine organisms don't need the heavy skeletons and other structures that support land plants and animals. Also, shapes mislead: the pale, graceful little 'plants' on the coral wall are actually animals – sea-fan hydroids. The multi-tentacled 'worm' foraging over the coral is as much a mollusc as is the garden slug; it has stolen the stinging cells of its prey and incorporated them into its tentacles for its own defence. The pieces of fine white cloth clinging to the sharp edge of the reef are not rags, but animals – ribbon bryozoans.

At less than a depth of 100 metres or so, there is enough light to support photosynthesis, the process by which plants convert sunlight and atmospheric gases into organic materials for growth. Here, microscopic free-floating plants, the phytoplankton, proliferate. These form the base of marine food chains, directly or indirectly supporting every one of the sea's creatures. The energy harnessed by plant plankton is passed on to the many tiny animals that prey on them, the so-called zooplankton, including minute shrimps, sticky-armed ctenophorans and innumerable marine larvae. Zooplankton are, in turn, the staple diet of filter-feeding creatures, from corals, sponges and fish fry to the enormous whale shark and the great baleen whales themselves, like the humpback and the blue.

There is also an amazing array of secondary consumers, including brilliantly coloured parrotfish, which crush corals like biscuits in their beak-like teeth, and the crown-of-thorns starfish, which extrudes its stomach through its mouth to envelop and digest coral polyps. Higher up the predator hierarchy, there are moray eels that writhe along the sea bed by night, and top predators like the killer whale, the great barracuda and the requiem sharks. Still other creatures target the dead or the dying, contributing to the dramatic rate at which life in the ocean is recycled. Unconsumed dead plankton and animal carcasses become drifting and sedimenting resources for foragers like heart-urchins, sea cucumbers and shrimps.

The remaining debris drifts downwards five kilometres or more, where it becomes food for a vast number of deep-living species. Crawling and skimming, burrowing and flitting on the muddy silt of the abyss floor lives a community of scavengers which includes brittle-stars, sea spiders, crabs, polychaete worms, nematodes, giant isopods, slime eels and sleeper sharks. A large animal that eventually plunges, dead, into the abyssal ooze, will be swarming with isopods within minutes, their feeding a race before slime eels arrive to twist away chunks of meat, and before the sharks come, drawn by taint in the water.

In places there are rocky peaks that rise up from the ocean floor, where great numbers of oceanic fish congregate. The rocks are often surrounded by swirling currents and, where a diver has to cling on or be swept away, the fish often seem to just hang in the water, making casual adjustments with fin and tail. Elsewhere there are upwellings, where bitterly cold deep water sweeps up to the surface, bringing with it some of the accumulated sediments from the sea bed. Much of this can be eaten, so upwellings support immense shoals of fish of all sizes, patrolled by hunters and sometimes ploughed by the feeding frenzy of high-speed predators like tuna. There are often fishing boats too, which belong to the supreme predator of the seas.

OCEANS AND HEAT

The oceans have many other parts to play in the biosphere. For one thing, they shift heat around. As we saw in Chapter 1, water has an immense capacity to absorb heat energy, requiring more heat to raise its temperature than almost any other substance, and more heat has to be lost before it can cool down. Under the hot tropical sun, the ocean surface warms up, storing heat, and then it moves away in the direction of the colder poles. Partly it moves because the Earth spins, and partly because of the tides caused by the moon's gravity. But mainly it moves because it's pulled. This works because every part of the ocean is linked to every other part, by the hydrogen bonds that connect water molecules with each other.

Warm surface water stays near the surface because it's less dense than the colder water underneath – in other words, it floats. But if surface water approaches the poles, it cools, and becomes denser and therefore less buoyant. Eventually it freezes, at minus 1.9°C, separating through 'brine rejection' into floating fresh water ice and salt, which remains dissolved in the water beneath the ice. This makes the surface water saltier than it was before, and therefore denser than deeper water, so it sinks. The combined effect of cooling and brine rejection means that an immense amount of cold, salty water vanishes into the depths, and into its place is pulled an equally immense amount of warm surface water from the tropics.

The cold water then flows back in the general direction of the tropics where, eventually, it will rise up against a rocky sea mount or a continental shelf and re-enter the surface waters. This doesn't necessarily happen in the same ocean where it was first warmed, though, since there are deep cold currents that edge their way around the continents. But within each ocean, once the dynamic of moving surface water is established (and if you add the momentum of an astronomical tonnage of water in motion), you get a stable, warm ocean current. The lands near the current's passage get warmed, since moving air picks up some of the heat and blows it onshore. Such ocean currents are the main force that distributes heat around the world, giving the surface of the Earth a far more even temperature than would be the case without them.

The classic example of this is the Gulf Stream, which for millennia has brought heat from the tropical Atlantic, past the Gulf of Mexico, and all the way up and across the Atlantic, past Europe, before cooling and sinking off Greenland. This has made maritime Europe much warmer than it would otherwise be. What the Gulf Stream gives, however, it may also take away, and global warming could upset the whole system. This would happen if the Greenland ice cap were to melt in a big way and suddenly, since that would put vast amounts of fresh water into the sea just where it is supposed to be sinking because of its saltiness. Diluting salty water reduces its density, so the northern end of the Gulf Stream wouldn't sink as fast or as hard, and far less warm water from the south would be sucked north. This would give a cooler climate to the countries around the North Atlantic. However, unless there's a catastrophic acceleration of ice-sheet melting in Greenland, by the time it happens its effect may be masked by a general increase in temperature. The long, hot summers of recent years in Europe don't yet seem to add up to such a catastrophe for the ice, but this could change.

WILD WATER

The heat contained in the oceans largely drives the weather systems that affect life on land, providing the water in clouds and rain, and the energy to deliver wind and rain far inland.

Warm air rising over warm water creates low pressure at sea level, which pulls air inwards to replace it. It also sucks upwards the surface of the sea itself, so a storm system with a deep low-pressure zone in its heart can contain a huge mound of sea water. Such storms spin because of forces generated by the rotation of the Earth, and this spinning can accelerate, taking more and more energy from the warm sea, building a higher mound of water surrounded by walls of clouds hurtling ever faster. This can become a hurricane, cyclone or typhoon, a sprawling, spinning storm with huge waves that can drown ships. Such a typhoon will wander for a time, anxiously watched by all concerned, until it subsides over cooler water, or hits the coast.

The worst combination is when a hot, high-energy hurricane with extremely fast winds, a huge burden of cloud water, and a massive internal dome of sea water touches a settled coast. Although it will quickly slow down over land by shedding its energy, it does this by scraping against the land surface – i.e. against trees, houses, people, etc., all of which fall over and fly about, often in fragments. I remember creeping in the lee of a blockhouse-like hotel in Hainan, south China, during Hurricane Fred, and seeing palm fronds and a dog flying horizontally over the lawn. And during Cyclone Alice, in Darwin, Australia, the corrugated iron sheets of a thousand roofs became flying guillotines. As the typhoon loses energy, it sheds water in the form of torrential rain, blasted against walls so hard that it can penetrate breeze-blocks, bubbling the paint on the inside walls. This rain drenches hillsides, lubricating layers of earth and rock so that they begin to slip over one another, until million-tonne mudslides occur burying whole communities, as happened with Hurricane Mitch in Nicaragua.

Then, as the inner part of the storm approaches land, the dome of water, several metres high, crashes over the shore and bears inland. Much like a tsunami, this storm surge is no wind-blown wave with air behind it, but a wall of water with more water behind it, pushing. Very little that we humans make or build can withstand this kind of treatment, and very little does. It may be that the bedraggled survivors are then given a respite, at least from the wind and rain, as the clear centre of the storm passes overhead; but afterwards the whole cycle will start again, from the opposite direction as they are exposed to the other half of the spinning storm.

EL NIÑO

Ocean surfaces, currents and bodies of water can vary in the heat they contain, with major consequences for life on land. *El Niño* is the name given by Peruvian fishing people to a warming of surface waters in the eastern Pacific. They notice this because it ruins their fishery, and also starves the sea birds that make guano, which they sell for use as a fertiliser. What happens is that a cold upwelling fails, cutting off the supply of nutrient-rich deep water that normally feeds vast numbers of fish, especially relatives of the anchovy called *anchoveta*. In normal times, this upwelling replaces surface water that is blown westwards into the Pacific by strong 'trade' winds. These are caused by areas of low pressure in the far western Pacific, which suck in air. This air is warm and moist after being dragged across the whole equatorial Pacific, and feeds the high rainfall of South-East Asia, and hence the rainforests of the Malay Archipelago.

Why the trade winds sometimes fail is not clearly under-stood, but when they do the movement of surface water from South America dies away, and with it the upwelling. When that happens, plankton populations collapse and the whole food-web is stalled, with population crashes among *anchoveta* and the fish and sea birds that prey on them. The 1957/58 El Niño led to the starvation of about half of the thirty million guano-producing sea birds in the area, while the 1972/73 El Niño contributed to a fishery collapse from fourteen million to two million tonnes.

El Niño has turned out to be a far vaster phenomenon than originally thought, affecting most of the equatorial Pacific and beyond. Because they are marked by a switching in the intensity and direction of currents and winds in the vast

Pacific Ocean, El Niño events have knock-on impacts that extend over vast distances. They are associated with droughts in East Africa, northern India, South-East Asia, north-eastern Brazil and Australia, and with catastrophic flooding and hurricanes from Mozambique to the Gulf of Mexico. It has gradually become clear that the great oceanic-climatic cycle of which El Niño is part – the El Niño-Southern Oscillation – is the key driver of weather patterns in Australia, and profoundly affects wildlife ecology, from the breeding behaviour of bearded pigs in Borneo to the spawning and bleaching of corals in the Indian Ocean.

Not surprisingly, events on this scale have profound economic consequences too. The El Niño of 1997/98 caused floods in Kenya, for example, which did so much damage to transport and water supply infrastructure and the health sector that the country's GDP fell by 11 per cent. At the same time, in the Philippines drought linked to the El Niño starved agriculture of water for the dry season crop, and in Indonesia there were food shortages that contributed to the political instability that brought down a regime which had been in power since the mid-1960s. The same El Niño drought was also a key factor in the forest fires that swept large areas of Sumatra and Borneo, creating a choking haze that damaged health across the region. The World Meteorological Organization estimated that extreme weather associated with the 1997/98 event seriously affected the lives of 117 million people, causing over 21,000 deaths and more than half a million illnesses, while driving nearly five million from their homes.

Other El Niño events in 2002/03 and 2006/07 have driven home the message that however well adapted we may be to 'normal' conditions, we remain vulnerable to extreme and unpredictable phenomena. And over all this is the increasing suspicion that El Niño events are becoming more frequent and more extreme with global warming. There is much debate about this, with the 2006 Human Development Report by the UN Development Programme describing the El Niño phenomenon as 'one of the largest – and most threatening – unknowns in climate change scenarios'. But the report does claim certainty in that:

> The incidence of extreme weather events is increasing, along with the number of people affected by them. During the 1990s an average of 200 million people a year from developing countries were affected by climate-related disasters and about a million or so from developed countries. Injury, death and loss of assets, income and employment from these events undermine the efforts of communities and governments to improve human development. Inevitably, the adverse impacts are greatest for people with the most limited resources. Since 2000 the growth rate in the number of people affected by climate-related disasters has doubled. Attribution may be uncertain – but there is at the very least a strong probability that global warming is implicated.

PROTECTING THE COASTS

One of the fears about global warming is that warmer oceans will spawn more severe storms. This, helped by lessons learned after the great tsunami of 2004, has led to much new thinking about exactly how storm surges affect people living in the densely settled coastal zones of the world, and how these impacts can be modified and their worst consequences avoided. For it *is* possible to design coastal landscapes in ways that make them less vulnerable to such disasters. One set of rather 'hard' approaches may involve pouring immense amounts of concrete or piling rocks into great sea walls, while also ordering people to live further inland outside a setback or no-build zone. Another, 'softer' set of approaches focuses on encouraging people to think through their vulnerabilities and take a mixture of measures to reduce them. These might include rebuilding coastal sand dunes by making traps for wind-blown sand, or planting coastal trees that contribute valuable harvests while increasing environmental security, or re-establishing wetlands where fish can breed and mature for catching later, and which can absorb sudden influxes of wave-water.

The whole issue of life on the coasts of a warmer world, though, is made that much more difficult because the sea itself is getting deeper due to thermal expansion and ice melt. Hence not only do the waves start closer to where people live and farm, but also the corrosive and erosive actions of wild water are delving deeper into what was once dry land. Sea salt is blown hundreds of metres inland, tainting soils and crops, while a less visible process is also underway underground. Here, sea water increasingly swills against the margins of rocks that bear fresh water, which feed the wells and springs that support life inland. If these aquifers are having too much fresh water taken out of them, which is common in islands such as the Maldives, and near many coasts from Indonesia to Mexico, then salt water will enter instead. Not so slowly, the waters beneath coastal lands are being made salty by this sea water intrusion.

POLLUTING THE SEAS

Not all the traffic is from the sea to the land. Immense plumes of dust are seen on satellite images stretching from the African Sahara Desert, for example, out over the Atlantic and curling round and over the British Isles, the North Sea and as far as Denmark. Or from the Central Asian Gobi Desert as far as Japan and the western Pacific. Or from the South American Atacama Desert into the south-eastern Pacific. These dusts deliver vast amounts of solid matter as well as chemical nutrients to marine ecosystems. More sinister is the creeping brown silt, eroded by rain from deforested hills far inland, that stains the surface of the sea in many areas. These silt patches stretch and bend along the coasts, drastically cutting down the light that penetrates the water and so undermining the ability of aquatic plants to photosynthesise. All too often, they also impose a suffocating muddy shroud over what were once living coral reefs.

Further out to sea, where we once thought the ocean was invulnerable, we are now finding oxygen-starved 'dead zones',

150 of them at the last count, and up to tens of thousands of square kilometres in area. Here marine algae have been fertilised to death by nitrogen and phosphorus compounds from agricultural fertilisers, vehicle fumes, factory emissions, sewage and other wastes. These cause blooms of phytoplankton, the rapid growth and decomposition of which uses up oxygen in the sea water, thus suffocating all other life and then, not much later, the phytoplankton themselves. Such dead zones were first found in Chesapeake Bay between Maryland and Virginia in the US, the Baltic Sea between Poland and Sweden, the Kattegat between Denmark and Sweden, the Black Sea between Turkey and Ukraine, the Gulf of Mexico, and in the northern Adriatic Sea between Italy and Croatia. Then they were found in Scandinavian fjords, and now they are also off South America, China, Japan, south-eastern Australia and New Zealand.

Silting and dead zone effects are both likely to increase with higher rainfall in a warmer world, and will strike at the core productivity of oceanic ecosystems. Meanwhile, much wealth has been obtained by individuals, and many costs avoided by urban and rural citizens, and companies, through the free dumping of wastes into the sea. These wastes have included raw sewage and concentrated sewage sludge from urban areas, fertiliser and pesticide runoffs from farmlands, eroded soil from badly logged forest catchments, and all manner of chemical by-products, clinkers, smelts and smokes from industry. The effects are often immediately obvious at a local level, with filth washed up on beaches or outbreaks of seafood poisoning. But perhaps the real worry lies in subtle impacts that are only gradually becoming apparent over huge areas. As well as impacts on climate, sea level and the quality of sunlight because of ozone depletion, these more subtle impacts take such forms as the toxins accumulated in sea birds that have never approached settled lands, in deformed and feminised fish, and in cancer-ridden dolphins. Meanwhile, baby Wilson's storm petrels are starving in the Antarctic because they are being fed fragments of plastic garbage by their confused parents.

GARBAGE PATCH DOLLS

It's not just the southern ocean that's now a rubbish dump. The North Pacific Gyre is a swirling vortex of ocean currents comprising 34 million km of the northern Pacific, between the equator and 50° north latitude. It's swirled by the clockwise pattern of ocean currents around it, the North Pacific Current to the north, the California Current to the east, the North Equatorial Current to the south and the Kuroshio Current to the west. The centre of the gyre is relatively stationary, but the circular rotation around it draws in floating material from much of the Pacific basin. This was once coconuts, palm fronds, logs and the wooden hulks of drowned fishing canoes, the larger items each with a dense cluster of fish sheltering beneath them in the years before they eventually decayed and broke apart. But since the birth of the plastic era in the 1950s, this biodegradable debris has been massively supplemented by plastic.

The central gyre is now a rubbish field as big as Texas, of floating bags, bins, bottles and buckets, flip-flops, dolls, yoghurt pots and polystyrene, fishing nets, nylon lines and floats, wastes that have caused mariners to rename the central gyre as the 'great Pacific garbage patch'. Rather than biodegrading through the action of living things, many of these plastics are only slowly broken down by sunlight, and as they do they become ever-smaller fragments of the same material. At all sizes they are indigestible and useless or worse to life, but smaller fragments are eaten nevertheless, often by jellyfish whose transparent tissues become spangled with internal debris. These then enter the oceanic food chain, ending up in the stomachs of fish, turtles, sea birds and dolphins. And not just as tiny items of clutter, either, but often as things like bottle caps, cigarette lighters and tampon applicators that, to a foraging bird, resemble food. More than a million sea birds, 100,000 marine mammals and countless fish die in the north Pacific each year, either from eating rubbish or by being snared by it. There's also the problem that small plastic fragments can attract and concentrate persistent organic pollutants, thereby becoming toxic as well as looking, perhaps, like extremely edible fish eggs.

Garbage enters the gyre from the coastal cities of East Asia and western America, and also as jetsam (items thrown away) or flotsam (items washed away) from ships. One case involved 80,000 Nike shoes, which became flotsam after five containers washed from the *Hansa Carrier* during a storm in 1990, south of Alaska. The shoes each had a unique serial number, allowing them to be identified as they floated around the gyre and were washed up on beaches around the Pacific. Another such case involved 29,000 plastic ducks, turtles, beavers and frogs, which became flotsam in a storm in the eastern Pacific in 1992. Many of these bathroom toys are presumably now in the great Pacific garbage patch, but some floated south, eventually washing ashore in Australia, Indonesia and South America. Others went north, some of them escaping the Pacific entirely, through the Bering Strait into the Arctic ocean, and after being trapped and released by ice, they've since wandered into the Atlantic. They floated south along the eastern seaboard of the USA, picked up the Gulf Stream, and are expected to reach the shores of Britain during 2007. These great, unplanned flotsam experiments have shed much light on the circulation patterns of the world's oceans.

THE SUPREME PREDATOR

Tropical waters are not very productive by world standards, because of their relatively low nutrient content and scarce oxygen. Nevertheless, the abundance and diversity of coastal marine life must have been very great in the virgin seas used by our ancestors, during the tens of thousands of years that they foraged their way along the warm coasts of the Old World. How much of this they could have harvested would have depended on the development of their brains, language, group co-ordination, skills and tools. Even in the early days, though, coastal waters would have offered enough shellfish to pick up, club open and satisfy hunger without much effort. Swimming, diving and rafting would have expanded their foraging area, as well as making short trips possible between islands, thus bridging sea gaps all the way from Africa to Australia.

During all this time, the vast bulk of oceanic productivity would have been far out to sea, or far away in colder waters – in any case, far out of reach. When ocean-going craft were finally invented and used to explore the world, false-starting in China about 600 years ago with the voyages of Zheng He (Cheng Ho), and then developing more permanently from a European base, the mariners were astonished at the vast fish populations they found, and the abundant predators that lived on them. The idea that the wildlife of the seas offered an inexhaustible resource for salting, drying, boiling down for oil, and trading for great riches caught hold and persists to this day.

With such an incentive, millions committed themselves to a life at sea, and fishing and hunting technologies steadily improved. Some marine mammals began to decline, including the larger whales during the nineteenth and early twentieth centuries: the Steller's sea cow was an early casualty, becoming extinct in about 1768. Meanwhile, the more accessible fish populations, such as cod in the North Sea, began to decline, especially after the introduction of steam trawlers in the 1860s. These were able to drag large nets across the sea bed, stirring up and catching every living thing larger than the mesh size. Signs of over-fishing became obvious soon after, at least to fishermen, but prevailing scientific opinion remained that these could only be localised and temporary impacts, given the size of the ocean and its fish populations.

The First World War put a stop to fishing in the North Sea between 1914 and 1918, and there were bumper catches in the years immediately after the peace, but these then declined again. Some saw this as a sign that there had been over-fishing in the pre-war years. Much the same happened after the Second World War, which suspended fishing from 1939 to 1945, but by then the international community had begun to seek ways to regulate and manage fishing. Negotiations for a comprehensive fishery management and conservation treaty among European countries were frustrated by conflicting national demands, although they did agree to limit mesh sizes in 1946. Meanwhile, the UN Food and Agriculture Organization (FAO) had been established, under the slogan *Fiat Panis* ('Make Bread!'), with the aim of ensuring rational use of the world's oceanic wildlife as a contribution to economic development. Its achievements on this were patchy, but FAO did at least manage to document the global fisheries disaster as it unfolded over the next fifty years.

At first it seemed not so bad, as there was a forty-year fishing boom, with annual catches rising four-fold between 1950 and 1989, before levelling off at about 90 million tonnes and then starting to decline. In the early 1950s, more than half of the world's potential fishing areas were barely being used, but within forty years almost everywhere was being fished to the limit, and today over a third of all fish stocks have collapsed or are in steep decline. The remaining stocks are being over-exploited to keep up the numbers, with diverse and unfamiliar species being marketed. This all adds up to a devastating cycle of over-fishing at the global level, although within the general pattern there are many smaller stories that could be told. In the 1990s, these included the collapse and closure of the Grand Banks cod population in Canadian waters, and the destruction of the orange roughy population in British waters, and in the 2000s, the loss of Napoleon wrasse from Indonesian coral reefs.

Catastrophic over-fishing worldwide is rooted in our trying to achieve 'rational' use, based on an inadequate understanding of wildlife populations and ecology. Advisory power was given to experts who performed calculations behind closed doors, influence was granted to stakeholders who controlled wealth and votes linked to fishing, and decisions were made by politicians and bureaucrats who were ignorant of marine ecology and in thrall to fishing lobbies. Meanwhile, the technological stakes were raised, in the form of sonar fish-finders, satellite navigation systems, targeted purse-seines (net curtains, hanging from floats on the surface and weighted at the bottom so that they can be drawn around a shoal of fish and then drawn tight to catch the entire shoal), factory ships and monofilament long-lines, so poor decisions would have dire effects before they could be corrected. And the global demand for fish products continued to increase with human population. All this bore down on marine ecosystems that had once seemed infinite and inexhaustible, but which have proved to be anything but.

MAXIMUM 'SUSTAINABLE' YIELD

To the extent that governments could exercise control over fishing, which is nil in international waters except through international treaty, and limited in coastal waters without major investment in policing, they tried to do so using scientific advice. This was based on principles of wildlife management, a science born when it was noted that deer and other terrestrial plant-eaters bred and grew faster once about half the adult population had been shot. The reason was that their food supply became more common when there were fewer mouths to feed. Thus, with hunting, many would die but the survivors would do better than they would have otherwise. It was soon understood that there must be some maximum rate of hunting that would not cause the target population to decline. This point was called the 'maximum sustainable yield', or MSY.

Further refinements were added to hunting on land, such as killing more males than females in harem-breeding species, since not all adult males would breed. Likewise, older and slower animals could be targeted in the cause of improving the average quality of the stock, an approach reflecting the idea that natural predators do prey populations a favour by weeding out the unfit. Since the hunters themselves were now the dominant predators, others, such as wolves, were killed as vermin. Finally, closed seasons were imposed to avoid disturbing the animals during the months of mating or calving, and young animals, and pregnant or milk-giving females were specially protected. By these means the MSY for target species was pushed to the limit.

All of this works well on land, where terrain can easily be owned, fenced and defended, where ecological conditions and the health and size of prey populations can be assessed by eye, where competitors and predators of the target species can be selectively removed, and where individual animals can be identified by age and sex before they are either killed or allowed to live. But in the sea, owning fish, counting fish, sexing and ageing fish, and selecting fish to catch are all much harder than the equivalent tasks on land, and far less is known about fish behaviour and the ecology of the sea. Nevertheless, the principles of MSY were optimistically applied to fish populations. Incredibly high harvesting quotas were allocated, and further inflated by fishery lobbying. For many fish species, such as cod, this was thought to be justified because adults produce huge numbers of eggs, so only a few would be needed to maintain the stock. In this vision, large numbers of older adults would feed the world, while their legacy of babies and youngsters would grow fast to catchable size amid surplus food and space, leaving a fresh crop of eggs behind when they were caught, and so all would be well.

There are many reasons why this was never going to work, including the following. Different species have very different growth rates and maximum ages, from a year or two for herring, to twenty or so for cod and upwards of a century for orange roughy. Environmental changes can kill variable numbers of eggs and young fish, so a large reserve of breeding adults is needed to ensure reproductive success across years. Fish species vary greatly in the number of offspring that reach maturity in ideal conditions, from a handful among sharks to hundreds among cod. Older adult fish often produce far more eggs each year than younger adults, for example among snappers, where a single ten-kilo fish can produce ten to a hundred times more eggs than ten one-kilo fish can in total. Some widely used fishing methods, such as trawling, do great damage to sea-bed ecosystems where fish breed, feed, shelter and mature, thus undermining the productivity of the ocean above them. And finally, many species respond to a low density of adults by clumping together, making the last ones easier to catch, and by breeding less, rather than more as expected. It's almost as if they lose the will to live.

THE WEB OF LIFE

Another flaw in the MSY-based management of fish popula-tions is ignorance, both of the fish themselves and of the web of life that they inhabit. This web is itself under intense pressure, from land-based pollution and siltation, and from a style of fishing that kills 30 million tonnes of other wildlife each year, over and above the 90 million tonnes of fish caught deliberately.

The scale of this 'web of life' problem is shown in a paper summarised in Charles Clover's book on over-fishing, *The End of the Line*. 'The paper, by E.V. Romanov, was an eye-opener,' he writes, since:

> Romanov estimates the catches of tuna in the whole western Indian Ocean at 215,000–285,000 tons between 1990 and 1995 – less [per year] than it is now. In the process of catching this, purse-seiners also caught up to the following amounts: 2,300 tons of pelagic sharks, 1,700 tons of rainbow runners, 1,650 tons of dolphin fish, 1,200 tons of triggerfish, 270 tons of wahoo, 200 tons of billfishes [sailfish, marlin and swordfish], 130 tons of mobula and manta rays, 80 tons of mackerel skad, 25 tons of barracudas, 160 tons of miscellaneous fish and an unspecified number of endangered turtles and whales.

At only about 3 per cent of the total catch, this random culling of the Indian Ocean's wildlife is actually quite limited compared with the overall rate of 25 per cent for worldwide fisheries. This is due to the targeted use of purse-seine nets in this case, rather than totally indiscriminate bottom trawls. Even so, little is known of the consequences of such targetless fishing for the oceanic ecosystem.

Despite all these problems with MSY, the idea continues to be applied in fisheries management, and is still what the international community uses to define best practice.

PRECAUTIONARY FISHING

Arising from the suspicion that the MSY approach will result in exhausted fisheries and a largely dead ocean, the main competitor to MSY is the 'precautionary principle'. Here, the idea is that if an act *might* cause irreversible damage to a public or private possession, such as a fish stock that could collapse for ever after over-fishing, then that act *must always* be justified and proven not to be harmful before being permitted. Thus, where an MSY system might start with a 50 per cent harvest, which will create an incentive to invest in fishing boats and gear by people who will then need to maintain the high quota, a precautionary one might begin with a 5 per cent harvest, followed by intense study of its effects. Only after the impacts of the initial harvest on all aspects of the fish population and its ecology are thoroughly understood, and a lack of harm has been demonstrated, might a proposal be entertained to increase the harvest.

The precautionary system is still expert-based, but it's far less likely to lead to over-fishing if the regulators have a degree of control over events at sea. And control is always going to be important, since the market for fish is effectively infinite relative to the number of fish, while current technology would allow almost every last fish to be caught, whether for human consumption or animal feed. So, in addition to precaution as a guiding principle, policing the seas will be essential, combined with restraints on using techniques that damage sea beds and non-target wildlife, and investment in studies of marine biology to find out exactly what's happening.

Few countries have yet adopted this particular combination, though Iceland and New Zealand have come closest by getting private companies to help with the costs of policing the seas and research. This is done by selling very long-term rights to a share of a more-or-less precautionary, species-based, national fish quota. The companies thus have a long-term interest in keeping the quota permanently high but sustainable, and therefore in co-operating with government to prevent over-fishing, and in providing data to improve the quality of decision-making. Although this would be improved by paying greater attention to non-target species and the ecosystem as a whole, we can recognise here a system than goes with the flow of human motivations, such as the quest for perpetual income security, and provides incentives to co-operate and understand the consequences of actions. Of course, there are draconian punishments in these systems as well, to deter cheating.

Arrangements like this seem to be the way to go for managing fisheries quite far out to sea, beyond about 10 km, where there are only a limited number of deep-water fishing craft at work, organised by a small number of companies. If such systems became widespread, the price of fish would increase as it became a luxury food, but at least there would still be fish to buy in the future. Which is fine for urban élites and wealthier countries, but what about the teeming millions of poor, small-scale fishers around the world who work close to shore?

COMMUNITY FISHING

Small-scale, 'artisanal' fishing is capable of damaging fish stocks by taking the larger adults and greatly reducing spawning rates. Even more serious impacts can also occur where everyone fishes by whatever means they like, especially when there is an influx of additional people. Such free-for-alls *have* depleted coastal fisheries in places, but seldom because of small craft using small nets or hook-and-line methods in mid-water, however numerous. Instead, the real problems come with ecosystem damage, with larger trawlers sweeping up and down the coast, dragging nets along the sea bed, by the use of explosives or cyanide to kill or stun fish to the surface, by *muro-ami* (trampling by people along a coral reef to drive fish into nets), or by the physical destruction or pollution of fish feeding and breeding grounds like reefs, wetlands, lagoons or mangrove swamps.

There's a growing movement for coastal communities to resist the destruction of their resource base, by seeking 'territorial use rights in fisheries', or TURF. The aim here is to re-establish ancient systems in which each community can make its own decisions on how to manage their environment and fish stocks. This is now common in the Philippines, where non-governmental organisations (NGOs) have contributed ecological knowledge to supplement traditional wisdom. It has also been shown that communities benefit by setting aside parts of their TURF areas as nature reserves, where the original ecosystem can regrow and fish can live long enough to reach their maximum spawning size, thus injecting larvae into the sea to colonise new areas. With TURF allowing the total fishing effort to be reduced and destructive fishing methods to be banned, and marine sanctuaries to encourage recovery, many communities have found that fish numbers and diversity have greatly increased.

THE VIEW FROM BELOW

The giant clams are doing well, here at Guiuan in Samar Island of the Philippines. Each is still anchored in its transplanting bed, the first arrivals 20 or 25 centimetres long, their iridescent, many-eyed mantles oozing from their heavy shells. There are more than a hundred of them now. Marge, their protectress, surfaces, blowing water from her snorkel, and slips onto the raft like a marine mammal. She looks around, noting the boundaries of the Guiuan community fish sanctuary, which runs along the beach in front of the coconuts up to *that* rock, then out to *that* islet, then across to *that* buoy with the sign on it facing seawards, then on to the eastern edge of the village. It's been five years since the community of Guiuan embraced the idea of her little charity, and made the fish sanctuary an official project of the municipality, using their devolved powers under the Philippines' new Local Government Code.

Back then, there was little down below the raft but black *Diadema* sea urchins and coral rubble. Most of the fish species that local people had harvested here were gone. The marine biologist remembers that one of her best arguments for the sanctuary was that the area was already useless, so it cost nothing to agree not to fish there. Now the ecosystem is rebuilding. Despite the dynamite and cyanide fishing, the trawling and coral dredging, and the diligent gleaning of living things from reefs throughout the Philippines, there had proved to be enough larvae out there to settle and colonise Guiuan's bare little patch. Soft corals and anemones arrived first, then some *Acropora* corals, then other stony corals, and crabs, mantis shrimp, myriad tiny reef fish – miniatures of angels, triggers, jacks, groupers, adult dottybacks and anthiases, the wriggling juveniles of harlequin sweetlips – and then finally the adults of bigger fish.

Nowadays the 'clam lady' can take old fishermen from the village to snorkel over the sanctuary, hearing their exclamations as they point excitedly to fish they never really expected to see again. Breeding in the sanctuary, the fish are starting to spread outwards. Fish catches are increasing even several kilometres away. Some tourists have come to stay for a day or two, since word of the giant clam recovery programme has spread on the Internet, with the help of some partner charities in Europe. She often brings visitors here from other communities as well. They go home to talk about the experiment at Guiuan, and fish sanctuaries are sprouting around the island. She is grateful for the new law, which made it legal for local people to control their own environments for their own benefit. This is a new thing in Samar, which since the Spanish arrived in the sixteenth century has experienced little but dispossession and rebellion. The biologist's people have experienced much defeat, but with their fish sanctuaries and community forests, they are starting to taste success.

The message is not yet universal, but in Aceh, Indonesia, there are some interesting cross-currents. Here, after the great tsunami of 2004, which destroyed large numbers of fishing boats, nets and other gear, several donors concentrated simply on replacing them. This would just restore the over-fishing of the pre-tsunami years, which had already reduced fish numbers by 70 per cent or so, changed the relative abundance of different species, reduced the average size of fishes caught, and increased the effort needed to catch anything. Other donors tried to restore coral reefs by transplanting corals at great cost, and by constructing artificial reefs of various kinds, including the use of massive concrete structures. But a very different approach was advocated by the UN FAO, based on the idea that fisheries and coral reef restoration should rely on natural processes for regrowth, in combination with the use of marine protected areas to provide sanctuary for spawning populations, while protecting the reefs from destructive fishing. It seems that the FAO has learned from its long observation of the global fishery disaster.

MARINE STEWARDSHIP – LIFE AND TAXES

Somehow we need to agree ways to bring pollution under control, to restore life to dead zones, and to rebuild coastal ecosystems that fend off sea-borne disasters. The crisis of the ocean has several different parts, each needing a different kind of action. As individuals, we barely interact with the deep ocean at all. The problems out there come from millions of tiny impacts by all of us, focused and concentrated through a few huge industries and government policies. Out there, solutions must involve laws that bind every country and through them, every corporation. It's quite feasible to implement such laws, now that technology makes the furthest part of the global ocean almost totally accessible. It's the negotiations that are the problem. The main need here is for political pressure on the negotiators by the public. People will also have to accept that policing international waters, and studying, monitoring and managing the life they contain, is going to cost money on a scale that can only really come from tax revenues.

The trick here is to make sure that taxes adequate to pay for policing the oceans are paid by those who use them, and in proportion to the benefits that they obtain. Everyone benefits a little from buying cheap fish, but the owners of factory trawlers and supermarkets benefit a lot from selling it. Thus, there is the intensely political task of working out who should pay what, complicated by the fact that no one really wants to pay anything at all. To create 'hard laws', ones that make people pay, needs a consensus that the time for the free-for-all is over, and the time for stability has arrived. In practical terms, this means 'show that you care': reward realistic negotiators who understand ecology with votes, letters, donations and resolutions of support from non-governmental groups. Meanwhile, get informed and stay informed, and exercise your rights as a chooser of competing products in the market place. The Marine Stewardship Council, for example, certifies certain products in the market as having been produced by sustainable fisheries. Buy them, and never buy the rest. This can make a difference.

COMMUNITIES AS PARTNERS

Then there are communities, and what you can do as a visitor to the coast. The most important thing is to remember where you are. The coast is special, and especially vulnerable. Every road and hotel room replaced part of its ecology, and every carload of visitors affects what is left, by making sewage and litter, by eating, drinking and walking the dog, and by spending money at holiday rates. These all have local impacts, and can affect local livelihoods dramatically. The solutions that local people are trying to reach deserve support. Every act by a visitor to the coast should help to strengthen rather than weaken local control of local ecosystems.

There are islands in Maluku (the Moluccas), in eastern Indonesia, where local people have a tradition of managing nature collectively. Called *sasi*, this was easily adapted to scuba divers, when the people realised that divers enjoy pristine coral reefs with plenty of fish, and are happy to pay for them. Divers spend at least a dollar for every minute that they spend underwater, if all their costs are considered, and often up to three if you include international travel. So a typical dive costs the diver up to US$150. This is only worth paying if the water is clean and the environment healthy. So the villagers did a deal with nearby dive resorts, to be paid the equivalent of a dollar a dive in exchange for their help in protecting the reefs near their homes. The money is given directly to the village, and has turned every single inhabitant into a reef guard, continually alert to the possibility of raids by dynamite or cyanide fishers from elsewhere. The result is that there are at least a few dive sites in South-East Asia that will still be worth diving in ten years' time. If every sport diver insisted on participating in an arrangement like this, then quite soon there would be tens of millions of people working to make the future safe for divers and the marine ecosystems that they visit.

Very similar ideas apply to coastal fishing and local pollution. Communities around the world are struggling for the right to manage their environments in their own way and in their own long-term interests. These are allies, and visitors should support them. If they are not struggling in this way, they should be, and visitors should encourage them. Sometimes they make mistakes, and visitors should promote the flow of information so that everyone can learn from everyone else. Every tourist is a bundle of global experience, just as every local person is a bundle of local experience. Visitors should get into the habit of talking a lot, and listening a lot, and questioning the arrangements that their tour companies and hotels have made on their behalf. It may not be everyone's idea of a holiday, but it will certainly help save the ocean.

Improvements are possible wherever people understand the ecological limits to harvesting and the links in their environment, and are able to defend their interests. Under these arrangements, especially if key fish-support areas like mangrove swamps are restored to health, fish productivity will increase over time, but the greatest benefits will go directly to the owners, managers and protectors of the fish and their habitats. Since communities are permanent, this mimics the long-term incentive structure offered by the Icelandic and New Zealand fishing models, where people *want* to help because it increases their own security.

Meanwhile, ocean water will carry on doing what it always has: nurturing life and making the weather, for good or ill.

5. SWAMP WATER

It was hot and dark that night in 2000, and the frogs were loud in the drainage ditches. The buildings and out-houses of the park headquarters in Pangkalanbun loomed black beside the deserted road. Suddenly there was a whiff of smoke, a red glow, a crackling noise, and the frogs fell silent. Figures fled as the fire began to roar, punctuated by explosions as it reached fuel stores, boats and vehicles. Within minutes, the offices, equipment and records of the Tanjung Puting National Park were no more than sparks and oily smoke, drifting and settling over a small town in Borneo. Within hours, there were parties in the logging camps and brothels on the main rivers in the park.

Then the workers went back to felling as many *ramin* trees as they could find in the swampy forests, scaring the orang-utans and hornbills with the buzzing whine of their chain saws. One by one the mighty trees would keel over into the dark waters, be stripped of their crowns and branches, and winched and floated out of the swamp. Eventually they'd find their way, with false papers, to sawmills as far away as Singapore, Malaysia and China, part of some 300,000 cubic metres of *ramin* logs stripped from the 4,150 km^2 park that year. This particular species makes a beautiful, fine-grained golden-yellow timber, more valuable than any other Indonesian wood. The loss of these trees would cause massive damage to up to 60 km² of the park in the course of 2000, savaging its most valuable lowland areas and orang-utan habitats, and again in 2001, in 2002, in 2003 . . .

GHOST DRAINS

In case you thought that logging was the only threat to the peat swamp forests of Borneo, think again, for logging is only a beginning. It's very damaging, of course, and opens up the forest to sunlight and hot, moving air, which dries it out. Many species adapted to ever-moist conditions in rainforest, such as filmy ferns, liverworts, leeches and amphibians, simply curl up and die in these conditions. But if that was *all* that happened, then many species would still be safe, either being able to move away from the chain saws, or recolonising the forest as it recovered and became moist again later. But peat swamp forests, which are widespread in Borneo, grow on waterlogged, organic-rich soil up to twenty metres deep, in which decomposition of fallen vegetation is extremely slow, because there's so little oxygen.

These swamps are supposed to be thoroughly wet, but loggers dig canals through them to gain access to the trees, and then to float the logs out. These canals drain the peat of water, and they are usually left flowing once the loggers have moved on. The result is *ghost drainage*, in which the whole peat bed bleeds out, and the soils and trees collapse. Then, with air in the peat, the agents of decomposition flourish, turning more than five million tonnes of dead leaves and wood per square kilometre into vast amounts of methane gas. This exhalation of decay, molecule for molecule, is twenty times more potent than carbon dioxide as a greenhouse gas. Much of the big wood in the fallen forest would take years to rot, but these days in Borneo fire often speeds things along. And it burns not just the vegetation but the very ground itself. The billions of tonnes of carbon stored in the peat pour into the air, along with vast amounts of acrid smog. The fires burn deeper and deeper for years, inextinguishable unless the drainage is blocked and the peat allowed to become water-logged again.

WHAT IS A WETLAND?

There's only one international treaty dealing with any particular kind of ecosystem, and it concerns wetlands. It is known as the Ramsar convention, since it was signed in 1971 in an Iranian town of that name. It defines wetlands to include swamps and marshes, lakes and rivers, wet grasslands and peatlands, oases, estuaries, deltas and tidal flats, near-shore marine areas, mangroves and coral reefs and human-made sites such as fish ponds, rice paddies, reservoirs and salt pans. The aim was to include land ecosystems that are strongly influenced by water, and aquatic ecosystems with special features due to their shallowness and closeness to land.

By this definition, wetlands cover some 12.8 million km^2, an area a third larger than the USA and half again as large as Brazil, or 2.5 per cent of the world's total surface area. Personally, I prefer to take out places that are permanently under sea water, thus losing the marine habitats, and also lakes and rivers, so as to focus on 'wet lands' – ground that's tidally, seasonally or occasionally under water or, if it's always under water, where it's shallow enough for ground-rooted vegetation to grow up through it. I'd also be inclined to lose the fish ponds, reservoirs and rice fields, although I'm happy to include areas that have been flooded to make a swampy nature reserve. But wetlands are zones of ebbing and flowing, expanding and contracting, flooding and drying, so there's no point in trying too hard to pin them down.

What we do know, though, is that wetlands are important. The mixing of soil and shallow water, with plenty of nutrients, oxygen and light, is usually enough to make the ecosystem very productive. From a human point of view, this makes for abundant harvests – of reeds, papyrus, wooden poles, fish, shellfish and water birds (and wildlife photos) –

with all that that implies for livelihoods and civilisations. Just as important, wetlands are often points of recharge for ground-water systems. They can absorb waters that would otherwise cause damaging floods, and their ecological complexity allows them to digest, recycle and make safe large amounts of sewage and other wastes.

Some of these values are locally temporary, since the water in a wetland may come and go, but each wetland may be a key link in a chain of life that spans continents. Many migrating bird populations, for example, touch down to feed in wetlands and could never complete their epic journeys without them. So conserving migrating birds often involves identifying and protecting the sites where they 'refuel'. Many of the 1,674 wetland sites designated by 154 countries under the Ramsar convention have the protection of migrating waterfowl as a key role.

This is not surprising, since the convention's full name is The Convention on Wetlands of International Importance, especially as Waterfowl Habitat. These Ramsar sites add up to an area of over 1,500 km², some of which overlap with the many thousand 'important bird areas' that have been identified by scientists working with the charity BirdLife International. One of the criteria used in this process was that they hold 'globally significant congregations of waterbirds, seabirds and/or migratory raptors or cranes'.

But birds are not the only migrants in and out of wetlands. In June 2007, experts from the Wildlife Conservation Society saw, in southern Sudan, parts of what may be the largest migration of land mammals on Earth, including an eighty-kilometre column of antelopes. Their aerial surveys confirmed the presence of an estimated 1.3 million of the local sub-species of kob, tsessebe and Thomson's gazelle, as well as 8,000 elephants, 13,000 reedbuck, 8,900 buffalo and almost 4,000 Nile lechwe, many concentrated in and around the Sudd swamp, the largest fresh water wetland in Africa. The animals are thought to migrate in response to changing water levels and food productivity across this whole ecological region.

WHAT IS A WETLAND WORTH?

The fact that wetlands have many different functions means that their economic role is often large. In working it out, economists try to give a monetary value to such services as flood control, storm buffering, recreation, water cleansing, biodiversity conservation, feeding and breeding of fish, water supply and wood or thatch harvests. Some important wetland services, like climate regulation and potential value to tourism, are often too hard to value and may be left out of the sums. This also applies to unknown wild species, which we know must be there and must have some value for just existing, but which tend to be ignored. Such undercounting can have an important effect on biodiversity conservation values, especially in tropical wetlands where most species are completely unknown to science.

Estimates of the total economic value of wetlands, per square kilometre per year, include US$113,000 for the Pantanal in Brazil, mostly from water supply, disturbance buffering, cultural values and waste treatment; US$8,774 for the Lake Chilwa wetland in Malawi, mostly from fishing; US$245,500 for the Muthurajawela marsh in Sri Lanka, mostly from flood control and waste-water treatment; US$863,000 for the Wadden Sea in the Netherlands, mostly from storage and recycling of organic matter and nutrients; US$95,750 for the Whangamarino wetland in New Zealand, mainly from preservation and recreation; and US$2.8 million for the Charles River wetlands in Massachusetts, mainly from flood protection, pollution reduction and recreation.

Looking at 89 such estimates allowed economists at the WWF and the Free University of Amsterdam to estimate the annual value of 630,000 km^2 of wetlands around the world at about US$3.4 billion per year. These areas were chosen because at least some information was available for them. Had they multiplied up from the Ramsar convention's estimate of 12.8 million km^2, they would have got a figure of about US$70 billion each year. In any case, they found that more than half the value of a typical wetland comes from ecological services (flood control, water purification and fish breeding), and almost all the rest from recreational use.

IN THE MANGROVE

A shady avenue of six-metre *Rhizophora* mangrove trees stretches into the distance, all clinging with stilt roots to either side of a mud dyke. A tidal canal, swarming with crabs, runs along the right side of the dyke as we head north, towards the Java Sea. To the left is a hectare or so of prawn pond, planted with mangrove trees across the middle, and completely lined by them. The dyke is punctured here and there by tubes that at high tide feed water and baby prawns from the canal to the pond. Each is guarded by a filter to keep out predators, but allows the prawns into their new habitat. Here they will grow until it's time for them to be harvested.

In every direction the landscape is the same: water swirling with life, ponds full of prawns, milkfish or seaweed, ditches, dykes and above all, curving bundles of stilt roots growing out of brown trunks surmounted by glossy dark green leaves drooping with the weight of hanging daggers. These narrow spikes are the propagules or fruit-roots of the trees: about 50 cm long, intensely sharp, maturing steadily in the Javanese warmth until ready to fall into mud or water, where they swiftly take root or else drift away on the tide to settle elsewhere.

There are three of us walking through this cool, green, watery environment in June 2006. Nyoman Suryadiputra leads us. He's head of the Indonesian Programme of Wetlands International, a charity that conserves and restores the parts of the world where water and land meet and mix. Petra Meijer, following, is Dutch, from a land mostly below sea level, and is based at the Malaysian office of the same organisation. Then there's me. My job is to find out, for the UN Environment Programme, how to re-create a mangrove swamp without moving all the people to the slums of Jakarta and letting nature take its course. Nyoman is showing us.

The local people had owned and occupied this landscape for many decades, living by fishing in the Java Sea. With the booming Indonesian economy of the 1980s and early 1990s, and increasing demand for prawns in the supermarkets of the world, the trend became established to clear mangroves for prawn ponds. This happened throughout South-East Asia, but Indonesia was particularly hard hit. The problem with this is that mangroves are incredibly productive ecosystems. Because they can tolerate conditions in salty mud, mangroves grow on tidal mudflats and along the fringes of lagoons and creeks in coastal areas. Here the mixing of nutrients from sea and land supports a seething mass of breeding and maturing fish, molluscs, sea cucumbers and crustaceans.

The result is that these swamp forests can yield an annual harvest per hectare of 100 kilos of fish, 20 of shrimp, 15 of crabmeat, 200 of mollusc and 40 of sea cucumber. More than seventy other uses for mangrove products have been documented worldwide, ranging from palm sugar and honey to tannin and water-resistant poles. For these reasons, mangroves help support the livelihoods of millions of coastal fisherfolk, and a decent area of such forest spreads its harvestable products far out to sea and up and down the coast. Thus, fishing peoples may depend on distant mangroves, even though they may not see the connection between them and the silvery masses of fish in their nets, or the prawns crowding around the lights of their traps at night.

Therefore, destroying a mangrove swamp to make a prawn pond privatises its productivity for the benefit of the pond-owner, but deprives many other people of their livelihoods. This took a while to be understood, by which time millions of hectares of healthy swamp had been destroyed, a few people had become rich, and ordinary fisherfolk were suffering badly (or had given up and moved to the cities). But the prawn ponds eventually proved unsustainable, because of disease and input prices, and many were abandoned, isolated from the tides, their mud oxidising to acid sulphates in the harsh sunlight.

TURN AROUND TIME

As recently as 1998, all the greenery and controlled, useful flooding that we were looking at eight years later, was just bare mud and stagnant water. The ponds had been artificially fertilised for years, and thrashed by mechanical aerators. But then the captive prawns began dying of the white rot, and because the economy had collapsed, no one could afford the chemicals and fuel any more. Then Nyoman's team chose the place for restoration. At the village of Desa Pesantran, they began working with a small group of men who were already thinking along the same lines, calling themselves *Mitra Bahari* ('Ocean Partners').

Not far along the coast, at Desa Nyamplung Sari, Wetlands International also started working with an all-women's group, *Bunga Melati* ('Jasmine Flower'), creating a kind of gender symmetry in the project. The women proved more effective at building businesses to use the products of the mangroves, but the men were better at the grubbier job of planting the mangroves themselves. Within a year of Wetlands International making contact with *Mitra Bahari*, its members were out enthusiastically tending *Rhizophora* seedlings in nurseries, growing them to the 'four-leaf' stage, and planting them around and inside their ponds. Every one of those hundreds of thousands of new trees has a place in the record book of the village group, maintained day-by-day back to 17 December 1999.

There is great vigour in *Rhizophora*, and if you plant them along both sides of a dyke, within a year or two not only will the banks be hardened against erosion (and so will no longer need much maintenance), but the stilt roots will quickly close off the walk-way between the lines of trees. Fortunately lopping them off seems to do no harm, and this is what people do. As the mangroves thrived, the local people became safer from storms and poverty, making a good living selling prawns, fish and seaweed, and carefully tending their trees, canals and traps. The resulting landscape is not a natural mangrove swamp, of course, but it *is* a productive and sustainable environment supporting thousands of people (and fruit-bats, herons, etc.). No wonder Nyoman was pleased to show it off.

ENVIRONMENTAL INSECURITY

In western Indonesia, morning prayers were long over by 7.50 a.m. local time on 26 December 2004, and the daily businesses of buying and selling, fishing, farming and preparing food were well underway. The people of Banda Aceh, strongly Muslim, were not taking it easy that Sunday, and they expected a normal working day. What they got, though, nine minutes later, was a shattering earthquake, one of the strongest that this earthquake-prone region had ever experienced, that felled people and furniture and buildings. Over the next half hour or so, people began to recover, tending the injured, looking at the damage, starting to pick things up. Then, in the middle of town, cries began to be heard in the distance, accompanied by a grinding roar that quickly came closer.

People ran to look, and then turned to flee from a new horror: an inky-black surge of water and floating debris that poured towards them along the city streets, unstoppably, with merciless inevitability. Some clambered onto walls or balconies, some made it into mosques, but everywhere there were the small desperate tragedies of men, women and children knocked down, smashed against walls, torn by planks, crushed by floating, spinning vehicles. For the longest time, Banda Aceh became one with the sea, but at last the turbulent, confused, polluted ocean drained away. What it left behind was beyond comprehension, even beyond description – there are only fragmentary visions: of the harbour surging with a mass of glossy green palm fronds and brown bodies; of a wilderness of mud and brown rubble where once had stood the capital of the proud Acehnese people; of survivors scrabbling for their children or sobbing among the broken corpses.

And that was just the *start* of the disaster for the peoples of the Indian Ocean, for similar scenes unfolded in many places. From the tourist resorts of south Thailand to the fishing villages of Sri Lanka and onward to the coasts of Africa, the same deadly waves came rushing in, exploding across break-waters, breakfasters, labourers and train passengers, obliterating settlements and then dragging many of its takings out to sea, grinding the dead across the reefs and sea bed, scattering the debris of decades of development far and wide. As the scavengers feasted, the dreadful process of coming to terms with it all began. When a person has lost everyone they ever loved, every resource they ever used to make a living, every element of the familiar architecture of their lives, they look like the people we all saw on our televisions in the days after Christmas 2004.

In Aceh, the tsunami had roared uninhibited across a muddy coastal acreage of almost-treeless fish and prawn ponds, where there had once been mangroves, and hit coastal and estuarine settlements and ecosystems across virtually the whole coastline of north-western, northern and north-eastern Sumatra. It need not have been quite so bad. Even so close to the earthquake's epicentre, with towering wave heights, if the water had rolled across hundreds of metres of mangrove forest, rather than bare mud, its force would certainly have been much reduced. For mangroves are physically sturdy and complex, with stilt roots and other structures which help to absorb wave energy. This is why healthy mangrove ecosystems help moderate wind-driven waves, and are important in limiting coastal erosion and storm damage.

Many areas with mangroves suffered less in the 2004 tsunami than those without such protection, although the effect was clearest in places where the tsunami waves were 5–10 metres high. Further away from the epicentre, for example in Sri Lanka, where the waves were a more reasonable size, their energy was largely absorbed and dissipated in those places where natural lagoons, mangroves and beach-dune systems had survived decades of logging and sand mining. However in Aceh, most of the natural mangrove forests had been degraded or destroyed in the years before that fatal Boxing Day, and in many places the remaining thin stands of trees were devastated along with everything else.

PUTTING BACK THE PIECES

In the context of a massive earthquake that doomed coastal ecosystems by heaving them permanently above the high tide area, or tilting them beneath the sea, or smashing them with an irresistible tsunami, you may think that nothing can be done. Yet this is a very rare kind of event. Much more frequent are the fierce storms and wave surges that seem to be becoming more common in today's greenhouse world, building on a steadily rising sea level that is bringing far more people within range of a disturbed ocean. In these circumstances, restoring coastal ecosystems starts to make a lot of sense.

By December 2004, mangrove cover in the twelve countries most affected by the tsunami was well below half its original extent. With ever-more people, settlements, resorts and roads being packed into the coastal zone around the Indian Ocean, the sense of vulnerability and hazard was greatly increased by the tsunami. Now the hunt is on for effective ways to put back mangrove-protected landscapes that are inhabited and used by millions of people. Many international organisations and government agencies all realised this at the same time, and since the tsunami more than thirty million mangrove propagules have been shoved into the mud and sand of Aceh's shorelines.

Not a lot of them are still alive, though. The reason is that mangroves need care and attention. They need to be planted in the right places. They can be washed away if not protected from strong currents, or covered by sand, or die of sun-bake if left unshaded when too young. Predators can kill them – for instance crabs, which have to be decoyed away using bamboo stems, which they feel, consider inedible, and move away from. Seedlings need to be grown in protected nurseries for some months before being planted in their ecologically correct locations, and, above all, post-planting community care is needed for maximum seedling survival and full establishment.

In other words, local people are needed as active partners, not just as hirelings for public works, and they need to know and care about what they are being asked to do. This is where the 'Ocean Partners' and the 'Jasmine Flowers' of Java can help. Facilitated by Wetlands International, and funded by Spain and the UN Environment Programme, coastal people from Aceh have already been spending time with them, learning how to plant and care for mangroves, and returning to their homes bearing knowledge and inspiration, new enthusiasts for the swamps.

FISH IN THE TREES

Tropical rainforests are very lively places, which is to say that they are biologically rich, enormously complicated, manifestly luxuriant and filled with all manner of plants and animals. Rainforest communities never sleep. Their inhabitants know neither winters nor dry seasons, and nightfall is greeted only by a change in the species eating, mating, hunting, living and dying. These forests grow only in parts of the world where they are guaranteed a year-round, fairly constant temperature of 18–30°C, and an annual, evenly distributed rainfall of more than 2,500 mm, sometimes four or five times more.

Whether they're in Indonesia, Congo or Amazonia, in architecture they have obvious similarities. There are soaring trees, their trunks often buttressed; a lofty canopy of branches and leaves, with deep shade below; climbing lianas and palms; plants growing on other plants; a rather bare forest floor, with lumpy roots writhing across it, a dusting of fallen leaves, and scattered ferns and seedlings. Butterflies dance in rare shafts of sunlight from above, birds make exotic noises, and insects sizzle and click. This restless background is often overlaid by the rattling thunder of rain on distant leaves, and the dripping of water as it drains out of the canopy.

The combination of warmth and rain makes the air of the forest very humid, often saturated with water vapour. Mists shroud the trees as sun succeeds storm, rising up to form heavy clouds that pour out rain in their turn. The hothouse ambience allows plants and animals to relax their guard against cold and drought, but there is a hidden cost to the community as a whole. For a rainforest is almost always wet, and water creeps or runs across every surface, trickles among its roots, prying everywhere. Its work is to dissolve or erode, breaking things up and sucking their fragments out of the forest and down to the sea.

What the water will take, if it can, is the substance of the forest itself, its tissues, minerals and biochemicals, its food and its structure. But the forest is adapted to this pressure, which has been unrelenting for millennia. In the heat and the damp, its living things vie with the water and with each other for the chemicals needed for growth and reproduction. Nutrients are snatched from bodies, living and dead, and clawed back from solution before they are lost: all the forest's many lives are geared to this perpetual obligation. Bathed in water, at war with water, rain and forest are friendly foes, for with less rain, the forest would die.

Some rainforests are also swamps. One of the greatest annual floods on Earth happens every year around the Amazon River in Brazil. The timing varies north and south of the equator but is largely driven by snow melting in the Andes; around November the river and its tributaries start to rise, gradually reaching their maximum depth in June or so. By then, the area covered by water has expanded from a dry season low of 110,000 km^2 of permanent wetlands to a peak of over 350,000 km^2. At this peak, and for months either side of it, the trees stand in water ten metres or more deep. The flooded forest is known as the *várzea*, but it's not uniform. Some branches of the Amazon system contain silty, 'white' water, while others are tannin-stained 'black', and poor in nutrients. The kind of water that floods each area makes a big difference to its ecology, since the 'white' waters deliver vast amounts of sediment to fertilise the forest each year.

The trees in flooded areas have features that allow them to survive life in a seasonal swamp, including breathing roots so they can take in air above the water line, and buttress roots to brace them against currents. There are few plants on or near the forest floor, but with the arrival of the waters, grasses detach themselves and form floating mats, and giant water lilies float in from the river fringes. The fish of the Amazon are now free to forage for fallen fruits, seeds and small animals across the normally dry forest floor. Remarkably, since most Amazonian fish are carnivorous or descended from carnivorous ancestors, many have evolved relationships with trees and shrubs. They act as seed dispersers, taking over the role more usually taken by birds and monkeys in a tropical forest. A number of fish have teeth designed to process certain kinds of fruit. There's even a 'piranha tree' that's used by some of the 28 piranha species in the Amazon. Meanwhile, the young of the largest fresh-water fish in the world, the arapaima, hatched at the beginning of the flood, forage in the forest to begin their growth towards an adult weight of 200 kilos and a length of up to 3 metres. As this abundance of fish disperse, feed and breed among the trees, they are hunted by fresh-water dolphins, giant otters, cormorants and the small crocodilians called caimans.

In 'white' water areas, these extraordinary flooded forests have always been more densely settled by people than elsewhere in the Amazon, because of their high wetland productivity and regularly fertilised soils. The same abundance influenced settlers to found their capital, Manaus, and other towns, close to the *várzeas* and its fish and fertile land resources. As a result, the floodplain forests are among the most threatened of all ecosystems in South America, largely due to logging and forest clearing to make way for farms and ranches. There have also been negative impacts from large development projects, such as dams and roads, and dangerous pollution by the mercury used in gold mining. Commercial fishing, three-quarters of which depends on the relationship between the forest and the river in the Amazon floodplain, has added to the problem by reducing populations of target fish species. This affects local people around the rivers, who have among the highest rates of fish consumption in the world.

WETLANDS' END

Regardless of their true value, wetlands everywhere are under threat. They are being over-exploited or polluted, their waters are blocked by dams or diverted to irrigate farms and cities, and often they are deliberately drained to make way for other uses of the land, especially agriculture but also urban expansion. Too often, they are seen as wastelands, disease-ridden swamps of no use to society. But they may also be competed over by different interest groups, such as the farmers, the water-users, the waste-dumpers, the real-estate developers, the duck-hunters and the bird-watchers, as well as being overwhelmed by demand for space and resources by growing populations and expanding economies. Because of all this, more than half the world's original wetlands disappeared during the twentieth century, including 54 per cent in the USA since 1900, 67 per cent in France in 1900–93, and 55 per cent in the Netherlands in 1950–85 alone.

The fate of the Everglades in Florida is typical. Here, half the original wetlands are gone, and the remnants are polluted and slashed by canals and roads. Wading bird populations have collapsed, and 68 native species are threatened or endangered while alien species invade. About 2.5 km³ of water seeps away each year, while mercury and phosphorus contamination is widespread. Downstream, the escaping fresh water damages coastal estuaries, and coral communities are showing their stress by a ten-fold increase in diseases since 1980. As south Florida continues to boom – its population is expected to triple in decades – bringing damage to the Everglades under control is likely to be very hard and very expensive. It is nevertheless being attempted, through an US$8 billion plan to undo the damage and restore waters to the swamp.

Elsewhere, though, the process usually continues unopposed by well-funded restoration projects, and about 40,000 hectares of wetland are destroyed each year. In Senegal, Djoudj national park is threatened as the Senegal River is blocked, diverted and used for rice farming, and polluted by agrochemicals. In Uganda, the Lake George wetlands are threatened by pollution from copper and cobalt mines, as well as uncontrolled charcoal burning. In China, over 90 per cent of the wetland plains of the north-east have been drained and converted to farming, while pollution has degraded wetlands near to cities, especially along the Yangtze River, and all wetlands in the eastern provinces.

In Indonesia, Java has lost 70 per cent of its mangrove area, Sulawesi 49 per cent and Sumatra 36 per cent, and up to 12 million hectares of wetlands were destroyed before 1996, with the rate of loss increasing in later years, especially as peat swamp forests were logged, drained and burned. In Nepal, the floodplain grasslands of the Terai have been reduced to fragments by the farms of people resettled by government programmes, with further pressures from water diversion and overgrazing. In India, large stretches of mangrove forest have been severely degraded in almost all areas where they are found. Wetlands in Pakistan, which include mangrove forests, inland wetlands and the ecosystems of the Indus Delta, have been overwhelmed, along with the accelerating loss, fragmentation and degradation of all the country's natural habitats. And in Thailand, large areas of wetlands have fallen to rice fields and urban sprawl, and at least 35 per cent of the mangrove forests have been converted to shrimp ponds, salt pans and rice fields.

WETLAND CHOICES

The 'business as usual' options are to privatise wetlands for prawn ponds and rice fields, to use them as waste dumps, or to drain, dyke, dam, dredge, canalise and concrete them over, or, if they contain a lot of timber, log 'em flat and let 'em burn. The idea that a swamp might have some merit beyond a one-off use has proved a hard sell during the exponential expansion phase of the world's economy, and half of all wetlands have already paid the price. Yet it need not be like this.

Flooding disasters occur when seas or rivers take back their own, drowning the works that humans have erected upon lands that are by ancient precedent claimed by waters. We live in a turbulent world, with hotter and higher oceans creating fiercer storms, and a warmer atmosphere bearing more water vapour to dump harder rain in unpredictable locations. In this context, the capacity of wetlands to absorb and dissipate water and energy in times of crisis is increasingly valuable. Environmental and livelihood security surely demand much greater caution over their management.

The Great Tsunami of 2004, and the calamitous hurricanes and storm surges of recent years, Hurricane Katrina among them, serve as reminders of our vulnerability when we pack vast numbers of people and huge amounts of infrastructure into coastal zones and floodplains. These events should prompt an urgent review of wetland management. Loss of human life on such dramatic scale may be the driver to change that benefits the whole web of life.

In short, we could accept what the environmental economists are telling us, and with it accept the common ownership and public value of wetlands, for fisheries and for their roles in waste digestion, flood control and disaster proofing. We could restore damaged wetlands and build sustainable local businesses using harvests and services from them. And we could raise local awareness of how wetlands and flooded forests provide environmental security. Then we could look to the future with confidence that, by removing the threat to wetlands, they will help remove threats to us.

6. LAKE WATER

THE KIDNEY OF NORTH CHINA

Old Hu lifted his best cormorant from her cage, slipping the ring over the bird's head and massaging it down her long neck. She was used to this treatment, and made no trouble with her dagger-like beak. Old Hu tied the end of the tether-line to the bird's ring, and made sure the other end was attached to the boat. The paraffin pressure-lamp hissed, its bright light gleaming off the soupy, blue-green water of Lake Baiyangdian. Old Hu stood up slowly and raised the cormorant on his outstretched arm, where she shuffled her wings and scanned the water. There wasn't much to see.

There hadn't been much to see here for years, on this 366 km² lake, the largest in North China – the 'Kidney of North China' according to the Party officials, who say that it purifies the land's water, or the 'Pearl of North China' if the ancient texts and tourism promoters were to be believed. Old Hu remembered when it was much bigger, and much cleaner, before the cities took so much water for reservoirs, and began to flood the lake with industrial wastes and sewage. Some young comrades from the Agricultural University of Hebei in Baoding had passed through Old Hu's village the other day.

They stayed long enough for a beer in Liwei's coffee shop, and to be overheard using unfamiliar, foreign words, like 'cadmium', 'zinc' and 'lead'. Old Hu thought about that for a bit.

Then there was a swirl of water under the hissing lamp, and Old Hu ordered the cormorant into action. Without hesitation, she darted and dived cleanly into the water, vanished for a moment and then bobbed up with a fish in her beak. Still in the water, she flipped its squirming body to get it pointed head first down her throat, and tried to swallow. The fish got as far as the ring, and she began to gag. Old Hu pulled her in with the tether, until he could lift her, still trying to get the fish down, into the boat. Then he yanked the fish out and threw it, still twitching, into a basket. The cormorant glanced at him reproachfully, and began to preen. He looked at the fish. Its rear end was twisted, so the tail looked all wrong; there were lumps and white patches on its body, and it only had one eye. As he examined it, the eye glazed. Old Hu remembered the time when he would've done something else with that fish than cook it at home. But there really wasn't much choice, these days. He picked up the cormorant again.

JEWELS IN THE CRUST

Any large and permanent body of fresh water is a very precious thing, with all sorts of uses and reasons to be valued, truly a jewel in the crust of the Earth. A total of about 90,000 km^3 of the world's fresh water lies within some five million lakes at any one time. This adds up to a formidable part of the biosphere and a huge resource for people to use, and not all of these uses cause problems for people or for nature.

Lakes form where the water draining off an area of land is blocked by something, whether a shelf of hard rock that makes ground water overflow onto the surface, or hilly terrain that denies a river its path to the sea, until it gets deep enough to find a new one. Artificial lakes, where a river is blocked by people, will be looked at in the next chapter, but all lakes have drainage basins or catchments – the area where rain is destined to flow into the lake. Anything washed off or leached from the land in the basin therefore finds its way into the lake, whether this is silt from an eroding landscape, salts from underground rocks, or agricultural chemicals from a farming zone. Lakes delay that water, giving whatever it contains time to fall as sediment to its bed, or to affect the living ecosystem of the lake itself. So a lake is like a sump, where all things stay for a while and where many of them end up.

Any lake, therefore, is subject to several processes, which may be in or out of balance, beneficial or destructive. The amount of water entering the lake relative to the amount escaping decides how big or deep it will be, how long its water must remain there, how much will evaporate, and how salty it will become as a result. These balances will change over time, seasonally or in the longer term, as rainfall alters in the catchment, making the lake shrink or expand, become saltier or fresher. Likewise, if rivers flowing into the lake are diverted for irrigation, then more will evaporate from farmland and less will reach the lake. If the irrigated fields are sprayed or fertilised, then these chemicals, or their derivatives, will also reach the lake, along with the reduced, concentrated water flow. If, on the other hand, rivers are diverted entirely out of the drainage basin, then the lake will dry up, fast.

And then, if there's a city on the lake shore, or on a river flowing into it, other factors kick in – sewage, for example, and the oily effluent of car washes, the soapy traces of household detergents, and the accumulated trash of street litter or escaped material from garbage dumps. Or, if there are boats and ferries plying the lake, there'll be diesel leaks and probably fishing too, with all its disturbance to the local web of life. Or maybe the authorities have decreed that new species be introduced to the lake, to 'improve' fishing. Then, maybe, there'll be cataclysmic ecological shifts as foreign catfish fight it out with native cichlids, or exotic crayfish battle indigenous clams.

However, it's important to see lakes as dynamic concentrations of benefits, options and opportunities, as well as sources of competition and conflict between the various uses that the divided minds of humans may dream up. Many lakes, for example, are home to immense populations of water birds, either permanently or during seasonal migrations. These can attract wildfowl enthusiasts, bird-watchers or bird-hunters, and can bring great financial benefits to the people who live around the lake, as well as encouraging more ecological and sympathetic management of the lake ecosystem itself. This chapter sketches what has happened to some lakes around the world, in evolutionary time and in human time, to illustrate the diversity of issues and challenges, values and wonders that these jewels in the Earth's crust can generate.

TOAD IN THE HOLE

Frogs and toads have thin skins that absorb salts, so they avoid salty water. When they are under water, they breathe through the same thin skins, but they must stay in water with plenty of oxygen in it. So in May 2007, when researchers spotted a toad foraging on the bottom of Loch Ness in Scotland, at a depth of almost 100 metres, they knew three things immediately: that the Loch is fresh, that its surface water must circulate to great depths, and that toads don't mind being under a pressure that would soon kill a scuba-diving human. I don't think any of them thought they'd solved the mystery of the Loch Ness Monster, though, or any other of the envisioned wonders of deep and remote lakes across the world. Besides, 'the Loch Ness Toad' lacks a certain something.

But lakes are full of wonders nevertheless, especially to a biologist. For they are islands in a sea of land, and like other islands their isolation can allow their animals and plants to adapt to local conditions without their breeding stock being mixed with other lineages. Over time, these local breeds will often become so different to their relatives that they become separate species, unique or *endemic* to that particular lake. The longer a lake is isolated from other waters, the longer evolution has to work on producing endemic species. Likewise, the larger a lake is, the greater its diversity of local conditions, predators and other challenges, and food sources and other opportunities, to which adaptation can occur.

THE BLUE EYE OF SIBERIA

At more than 25 million years, lakes don't come any older than Lake Baikal, in southern Siberian Russia. Its location is important, since it is too far south to have been scoured out by glaciers during past ice ages. Its geological setting is vital as well, since it's in a rift valley where the Earth's crust is gradually pulling apart. This is the deepest continental rift on the planet, and at 1.7 km Baikal is the deepest lake, holding 23,600 km^3 of water on top of no less than 7 km of sediment. Lake Baikal is surrounded by mountains from which flow about 300 rivers and streams, and it has one outlet, the Angara River. As befits a lake in a deep gorge, Baikal is long and narrow, and despite its great depth, its cold waters are mixed and well oxygenated right the way down. In this it's like Loch Ness, and although no toads have yet been seen on its bottom, there are reports of a Baikal Monster, said to be a giant sturgeon, that preys on endemic seals. These seals, the only lake-dwelling seal species in the world, are just one of over 1,240 endemic animals in Lake Baikal, which make it the world leader among lakes in the abundance of endemic species. Time and isolation have done their work well at Baikal, with – at present – only the effluent of the Baikalsk Pulp and Paper Mill on its shores to endanger the evolutionary result.

THE GREAT LAKES OF AFRICA

The African Rift Valley is part of a colossal structure that runs from Lebanon to Mozambique, dividing into eastern and western rifts in East Africa. Its great lakes, like Baikal, also contain untold wonders of life, although their tropical waters mix poorly so their depths have little oxygen (and no toads). Lake Tanganyika, with 19,000 km^3 of water, is the oldest at 15 million years, the deepest at 1.5 km, and has at least 632 endemic species. Lake Nyasa or Malawi holds 8,400 km^3, is over two million years old, and has even more fish species than Tanganyika, with 423 endemics. The variety of life in this lake is remarkable, particularly among the vast family known as cichlids – highly coloured, territorial and specialised fish. Lake Victoria is the big one of Africa, with a huge surface area and about 2,750 km^3 of water, but, being between the eastern and western rifts, it's barely 90 metres deep and less than 20,000 years old. Yet this shallow youngster of a lake still manages to harbour up to 500 endemic cichlids. Or it used to: the Nile perch was introduced in 1954 and ate more than half of them into extinction. Still, where these much-mourned cichlids came from is a puzzle, and the lead candidate is Lake Kivu. This is much more ancient and, like Tanganyika, lies in the western rift. It sometimes overflows when vast bubbles of methane erupt within it, washing seed stocks of cichlids downstream to Victoria. Even so, if Victoria was dry so recently, as it seems, and only a few cichlids came from Kivu, the adaptive, species-producing power of these fish must be truly awesome.

LAKE NAIVASHA, KENYA

Lake Naivasha is a shallow fresh water lake in the eastern rift, about 80 km north-west of Nairobi. Two rivers feed it, the Malewa and the Gilgil, but the lake has no natural outlet on the surface. Instead, water seeps out into the ground, giving enough flow to keep it fresh. It's an internationally valued wetland site, and an important bird area, with such globally threatened species as the grey-crested helmet-shrike, Basra reed warbler and lesser flamingo, and large congregations of red-knobbed coots, African spoonbills and little grebes finding shelter there. The acacia woodland that existed around the lake was the habitat of many large mammals, and *Elsamere* on the lake shore was once the base for naturalist Joy Adamson's lion-rearing activities, a place that she chose because of its outstanding natural beauty. All the land around the lake is privately owned, and there are a number of wildlife tour operations and two hotels on the shore.

The production of cut flowers and vegetables for export began around Lake Naivasha in the early 1980s and quickly developed into one of the top three foreign-currency earners in Kenya. The farms now take up about 4,000 hectares of land immediately around the lake, and they depend on pumping irrigation water from it, which has helped reduce its level by around 2.5 metres. Meanwhile, forest cover has declined in the wider catchment, greatly reducing flows in the Malewa and Gilgil rivers. Growing flowers and vegetables creates much employment, and the population around the lake has increased greatly, from 7,000 in 1969 to an estimated 300,000 now. This has increased demand for domestic water, and for other uses such as watering livestock and washing vehicles. Meanwhile, unplanned settlements have grown up, with many households not connected to the municipal sewage works, which itself hasn't worked properly for years. So sewage enters the lake in large amounts, and also contaminates the ground water around it through widely used pit latrines. Finally, large amounts of fertilisers and pesticides are applied daily to the fields, and find their way into the lake. In short, Lake Naivasha is steadily drying up, and being poisoned.

The drying process is probably at least as much to do with rainfall as with irrigation, since the lake has a history of vanishing after prolonged dry spells, being entirely dry and under farms at the beginning of the twentieth century, and with its area varying between a hundred and a thousand square kilometres over the following decades. But profound impacts have accompanied deforestation in the lake's catchment and the transformation of a beauty spot housing fewer than 10,000 into a centre of industrial horticulture, with nearly a third of a million inhabitants. This would have greatly affected the area's and its people's ability to adjust to rainfall change. That this was done without rational planning, with every enterprise for itself, also meant that conflicts inevitably arose among diverse stakeholders.

Demand for farming and living space competed with nature conservation, while the use of agrochemicals, driven by competition among growers, was in conflict with the interests of public health and lake ecology. Even the lake's fishery, which was based on introduced fish in the absence of native species, was being seriously over-exploited. Meanwhile, the ecosystem that supported the fish themselves was being altered by an aggressive crayfish species, which was introduced in 1970. There are now attempts to reconcile these various interests through dialogue around the concept of a Lake Naivasha Management Plan. But the degrees of freedom needed for a complete solution have already been so reduced that the challenges seem all but insurmountable.

To a greater or lesser extent, something similar can be said about all the other eastern rift lakes in Kenya. Lake Nakuru has been losing its flamingos, apparently because of waste-water dumping by nearby industries and increasingly wide variation in water inflow due to catchment deforestation. Lake Bogoria, a World Heritage Site, is threatened by agrochemical pollution and siltation from eroding, once-forested, farmlands. Lake Baringo, further north, is reduced by over-irrigation and is badly over-fished. And the water level of the northernmost lake, Lake Turkana, has been shockingly reduced since the 1970s, mainly due to irrigation and a hydropower dam upstream, combined with prolonged drought. The problems always seem to arise when there are several conflicting uses of lake ecosystems, with no single interest group in charge. This is especially so when a lake has been stressed by people and is then subjected to another environmental disturbance. In a warming world, and a drying Africa, we can expect to see this knock-out combination more often.

THE ARAL SEA

But there are many paths to perdition, if you're a lake, and one of the most spectacular ecological disasters of the twentieth century happened to a lake precisely because one interest group was in charge. This was the government of the USSR, and their decisions doomed the Aral Sea. At about 67,000 km^2, this was once the fourth-largest lake in the world, after the salty Caspian Sea and lakes Superior and Victoria, and contained over 1,000 km³ of fresh water. It had abundant fish resources, and a busy shipping trade between its northern port of Aralsk and the ports of tributary rivers. Then, in 1918, the new revolutionary government of Russia decided to divert the two rivers that feed the Aral Sea, the Syr Dar'ya in the north-east (in what is now Kazakhstan), and the Amu Dar'ya in the south (in what is now Uzbekistan and Turkmenistan). The Amu Dar'ya was once known as the Oxus, and was crossed by Alexander the Great in 329 BC to allow his conquest of Tajikistan and Afghanistan. The Soviets' intention towards the river was less warlike, but ultimately more destructive. It was to make it irrigate a huge area of desert, in order to grow rice, melons, cereals and, most importantly, cotton.

The concept was to create a cotton landscape in the Soviet republics around the Aral Sea, and the extent to which it succeeded is still visible in the inheritor states of the USSR, especially in Turkmenistan and Uzbekistan. These have the first and second highest per-person rate of water consumption in the world, not because they wash or drink a lot, but because they grow vast amounts of cotton, one of the world's thirstiest crops. The other former Soviet republics with access to diverted water in the Aral basin, Kazakhstan, Tajikistan and Kyrgyzstan, occupy another three of the top seven places for agricultural water use per person, all of them withdrawing more than 2,000 tonnes per person per year. By comparison, annual per-person water use in the USA is 1,800 tonnes, in France 650 and in the UK 200. The global average is for a tonne of finished cotton textile to have used almost 9,400 tonnes of fresh water in its production, or about 2.8 tonnes per T-shirt. You'd have to run a shower all day, from seven in the morning to six or seven in the evening, to use that much water. In Uzbekistan, though, it takes more than 11,300 tonnes of water, and in Turkmenistan over 15,000 tonnes, to make a tonne of cotton fabric, largely because they are so wasteful in using it. This water once filled the Aral Sea, but now, in effect, it's exported as cotton.

The irrigation canals began to be built on a large scale in the 1930s, but the water mostly remained within the Aral catchment. Here much of it evaporated, and the rest was polluted, but at least some of it eventually reached the lake. The Karakum Canal, however, was built in the late 1950s to take water from the Amu Dar'ya River for use far outside the lake's catchment, in southern Turkmenistan. By 1960, therefore, up to 50 km^3 of water were being diverted each year from the Aral Sea, and it began to die. From 1961 to 1970, its level fell an average of 20 cm per year. The canal was extended in the 1970s and 1980s, however, and as this happened the rate of the lake's decline increased, from 50–60 cm annually in the 1970s, to 80–90 cm in the 1980s. The Aral Sea lost more than half its area and over two-thirds of its volume, while greatly increasing its salinity as it evaporated. None of this came as a surprise to Soviet planners, who had simply and consciously chosen one use of water over another.

The ignored consequences for local people around the Aral Sea, depending as they did on farming, fishing and shipping, were of course disastrous. The receding sea left huge plains covered with salt and toxic chemicals, and about 200,000 tonnes of salt and dust are carried by the wind from the dry lake bed every day. This poisonous material falls on fields and destroys farms and pastures. Fishing ceased completely, and the fishing port of Moynaq in Uzbekistan, where 60,000 people once worked, now lies far from the shore. Meanwhile, shipping and other water-related activities drastically declined, as did agriculture, and rising unemployment led to a major exodus of people. The quality of drinking water has horribly declined due to salinity, bacterial contamination and the presence of pesticides and heavy metals. Anaemia, cancer and tuberculosis, and chronic allergies such as asthma, are on the rise. The incidence of typhoid fever, viral hepatitis, tuberculosis and throat cancer in the area is three times the national average.

In 1987, its continuing shrinkage split the lake into two separate bodies of water, forming the North Aral Sea in

Kazakhstan, and the South Aral Sea in Uzbekistan. A channel was cut to link them, but that connection was gone by 1999 when the surface area of the lake fell below 28,500 km. By 2004, it was only 17,160 km, barely a quarter of its original size, and still contracting. By that time, of course, the USSR had been history for more than a dozen years, and the countries in charge of the lake's basin were Uzbekistan, Turkmenistan and Kazakhstan. The Kazakh government, at least, seemed to take the future of the lake seriously, and efforts were made (with World Bank financing) to improve the flow of water in the Syr Dar'ya and to dam any leakage from the main part of the North Aral Sea. As a result, water levels in that part of the lake have started to increase, and its salinity to decrease. Economically significant stocks of fish have returned, and by 2006 catches were being exported again. The port city of Aralsk, meanwhile, was only 25 km from the water's edge, 75 km nearer than it had been. A moister microclimate was also being established, bringing rain and hope to farmers in the regional dustbowl that had been created. Whether any part of the South Aral Sea can be restored like this depends on decisions made in Uzbekistan and Turkmenistan.

These two countries have long divided and taken the Amu Dar'ya water to maintain and even expand the cotton landscape created by the Soviets. But times have become truly hard in Uzbekistan, especially near the remnants of the South Aral Sea in its once-fertile autonomous region of Karakalpak-stan. Here there are dust storms fifty days a year, bearing salt and pesticides in particles just the right size to breathe in. The soils are full of salt brought from underground by evaporating irrigation water, and 200,000 hectares of land have been abandoned. The rest must be washed before planting, using yet more water which is itself becoming scarcer with harsh and frequent droughts. Grinding poverty, a deteriorating climate, salt, pesticides and general ill health have reduced life expectancy to only 51 years.

Elsewhere in Uzbekistan things are little better, with half the fields salt-contaminated, half the water in the canals leaking into the ground, and half the remaining water used to wash salt back into drainage ditches where it's then used on farms downstream or, in very hard times, for drinking. As the cotton-based lifestyle becomes increasingly impossible to sustain, there are moves to re-create the livestock-based ways of living that prevailed before cotton arrived. Certainly, impoverished Uzbekistan may have an interest in following Kazakhstan in re-thinking its relationship with the waters that once fed the Aral Sea, and with the lake itself.

Matters, however, are very different in Turkmenistan, far to the south of the lake. Here the Soviet-era deal that awarded shares of water in the Karakum Canal is still maintained, for it gives 22 km^3 per year to the 5 million people of Turkmenistan, and the *same* amount to the 27 million of Uzbekistan. This inequality was amplified when enormous reserves of natural gas and oil were discovered in Turkmenistan, which then needed only a mad dictator to spend its riches with suitable extravagance. The country got one at independence, in the person of President-for-Life Saparmurat Niyazov. Vast and expensive projects soon followed, including palaces, monuments, public fountains and a huge artificial lake, along with a cult of personality and a police state. Niyazov ruled from 1991 to the end of 2006, when he died suddenly and was replaced, through an implausible election, by Gurbanguly Berdimuhammedow. The new regime has tended to focus its attention on managing profitable gas sales to Russia, rather than on ecological restoration in the Aral basin, but perhaps this too will come one day.

LAGUNA DE BAY, THE PHILIPPINES

Laguna de Bay is the largest lake in the Philippines, with the sprawling vastness of Metro Manila on its western shore and occupying much of its catchment. The lake's area is over 900 km^2, but its average depth is less than three metres so fish traps and pens can be built right across it. It drains to Manila Bay through the Pasig River, and is used in many ways, few of them compatible with each other. It's plied by passenger boats, provides food for people and livestock such as ducks and captive fish, and yields water for power stations and irrigation, as well as being a sump for domestic, agricultural and industrial wastes. But most of all, along with the Pasig River, it's a sewer.

Since 1997, when the municipal authorities privatised the water supply and sewerage in Manila, there's been a focus on increasing access to clean water. But sanitation has received far less attention, partly because of the huge scale of the challenge, and partly because of a legacy of under-investment in something that's almost invariably seen as a low priority until people actually start dying in the streets. Thus, less than 4 per cent of Metro Manila's population is connected to the sewer network. Richer households have built their own sanitation facilities, while housing developments are often connected to common septic tanks. Around 40 per cent of households now have on-site latrines, many with flush toilets, and there are at least a million septic tanks in the city. The problem is where the substances that are flushed end up. For there are few facilities for treating and disposing of sewage sludge, and it tends to be dumped instead into the complex network of waterways that links the lake to Manila Bay.

More than 60 per cent of a population of close to ten million discharge their wastes directly or indirectly into the lake. As a result, about 70 per cent of the Biological Oxygen Demand (a measure of organic filth content) in the lake's water comes from households, with another 20 per cent from industry and 10 per cent from land runoff or erosion. Meanwhile, hundreds of factories discharge a mixed array of pollutants, including hazardous chemicals containing lead, mercury, aluminium and cyanide. Rapid agricultural growth in other parts of the lake's catchment has involved massive use of fertilisers and pesticides, the residues of which find their way to the lake where they induce rapid algal growth, oxygen depletion and fish deaths. At least in the lake there's some dilution, but you can almost walk on the Pasig, one of the world's most polluted rivers, with faecal bacteria levels exceeding government standards by a thousand-fold or more.

Not surprisingly, a third of all illness in Manila is waterborne. And during the dry season, between November and May, the Pasig reverses direction and carries pollution into the lake itself. So Laguna de Bay is not in good shape, and none of the various blueprints for cleaning it up have ever left the drawing board. It does have spectacular sunsets, though.

LAKE TEXCOCO, MEXICO

There are worse fates than being polluted. At least Laguna de Bay is still a lake, but Lake Texcoco in the Valley of Mexico is almost entirely in the past tense. This was once a shallow, closed, salty lake, about 1,000 km² in area and surrounded by marshes and forests. Within it, in 1325, the Aztecs built their capital, Tenochtitlán, on a cluster of islets in the western part of the lake. They surrounded it with an artificial island, and built elaborate systems of dams, rain-traps and channels to control the lake's waters and provide drinking supplies. From here, they and their allies ran an empire in which semi-autonomous city-states paid tribute in the form of feathers, adorned suits, stone beads, cloth, firewood and food, usually organised by local nobles and produced or paid for by commoners. It was also a commercial empire, with trade and trade goods, a currency of cocoa beans, and established market prices for such conveniences as selling a girl as a religious sacrifice, for which her father could expect 500–700 beans. It was, therefore, to us a very alien society, but one with familiar features, including a feudal hierarchy with nobles, peasants, warriors, artisans and traders, and slaves at the bottom.

Meanwhile, around and beneath all this activity, there was Lake Texcoco. At first, near-perfect coexistence was the major theme, with the lake providing food, transportation and raw materials. Houses were made of wood, placing surrounding forests under pressure, but they were thatched with lake reeds. Around the island, fenced mud *chinampa* beds, fertilised with human excrement, were used to grow food, and gradually became incorporated into the island city as it grew. Supported by these highly productive beds, which could yield seven harvests a year, the population of Tenochtitlán and nearby cities on the lake shore grew to 700,000 people. Such a population would have had considerable ecological impact, but Texcoco was still a lake in 1521, when Spanish forces arrived along with an army from cities newly freed from Aztec domination. After a brief and bitter struggle, Tenochtitlán was conquered, and the Spanish founded Mexico City on its ruins.

During the siege of Tenochtitlán the city's protective dams were destroyed, and never afterwards rebuilt, so flooding became a problem for the new city. There were serious floods in 1604 and 1607, and from 1629 most of the city was awash for five years, despite earlier attempts to dig drains. The Spanish never really tried to live with the lake, as the Aztecs had done, and eventually decided to drain it. This they accomplished through channels and a tunnel to the Pánuco River, thus creating an outlet for Lake Texcoco for the first time. By 1864, only about 230 km^2 remained of the original lake, and it was further reduced to 95 km^2 by 1891. But by then, the city's underpinnings had dried and shrivelled and the city had sunk, making it even more prone to flooding. It was not until the 1960s that a deep network of tunnels finished the job, making the city almost immune to flooding, and deleting Lake Texcoco from the map.

The ecological consequences of the draining were enormous, and parts of the Valley of Mexico became extremely dry. Following an explosive growth in population, from 345,000 in 1900 to around 16 million today, water was increasingly pumped from underground, causing the city to sink by several centimetres each year. By the late 1960s Lake Texcoco had disappeared, and what was left of its bed in the eastern suburbs of the city was salty, dry and barren. It was also being used as a dump for urban wastes, which would infest dry-season dust storms with the spores of bacteria and other pathogens, causing outbreaks of disease. In 1971, the government decided to restore this last remnant of the lake bed, 11.6 km^2 in area, by creating artificial lakes and marshes. These wetlands have since become an important feeding ground for migrating birds, despite continuing to receive sewage, being roamed by feral dogs, frequented by wildfowl trappers, and constantly threatened by the growth of slums. But at least a plan to turn it into an airport has been shelved.

OTHER LAKES, OTHER ISSUES

This sample of lakes around the world, their associated catchments, wetlands and floodplains, and their fates and the lessons to be learned from them, could continue for hundreds of pages more. There are, after all, some five million lakes in the world. We could have looked at Lake Vostok, 5,300 km^3 of super-cooled fresh water, super-saturated with oxygen and sealed below three kilometres of Antarctic ice. Or Lake Ohrid in Macedonia, five million years old and rich in endemics across the whole food chain. Or Lake Titicaca, high in the Andes, where Bolivia and Peru sort-of co-operate and sort-of compete over the lake's water. Or Lake Chad, where four West African countries meet, and where dams and diversions of the Hadejia, Jama'are and Logone rivers have caused catastrophic drying of lake and floodplain. Or the five Great Lakes of the USA and Canada – Huron, Superior, Erie, Michigan and Ontario – together containing almost as much fresh water as Lake Baikal, but with more than half North America's heavy industry in their catchments. Or Iran's Lake Hamoun, desiccated by dams on the Helmand River in Afghanistan, thus creating 300,000 environmental refugees, covering a hundred villages in sand dunes, and wiping out a thriving fishery with an annual catch of around 12,000 tonnes. Or we could have looked at the disasters waiting to happen, such as the glacial lakes in the Himalayas, bloated with melt-water from global warming, and temporarily blocked by ice and rock barriers that are just waiting to burst. But instead we'll see what patterns have already emerged from this brief tour.

MANAGING LAKES: THE HARD WAY

The lake management we've seen has all been pretty dire. Alien species have been introduced to 'improve' lake fisheries, in ignorance and with the intention of correcting nature. The result has usually been calamitous for lake endemics and disruptive for the established lake ecology, sometimes to the detriment of fishing lake-dwellers as disturbed wildlife populations oscillate, trying to find a new balance. Meanwhile, powerful individuals have often decided that lakes, their ecosystems and all their dependent peoples, are worth less than the crops that could be produced elsewhere using their waters. The resulting reallocation and redesign of ecosystems has often had unimaginable costs for the weak.

MANAGING LAKES: THE CHAOTIC WAY

Other examples tell a different story, and here the prevailing theme is chaos. Lake Naivasha, for example, is at the mercy of competing private landowners, all with an incentive to maximise the production of flowers and vegetables, and minimise costs. This may be an effective business model, but it's not sustainable around a real-life ecosystem like a lake. Meanwhile, other stakeholders clamour for rights, or just take what they need: space for living, water for livestock, the lake for their wastes, wood from its catchments. Still others are trying to run tourism businesses, which need at least the semblance of a pristine environment to keep clients coming. Then again, Laguna de Bay is just too close to a vast and dysfunctional city, one that's flooded by the landless poor of an over-populated, deforested and impoverished country. Chronic under-investment in the collection and treatment of sewage, and too many informal settlers who exist below the radar screens of sanitation planners, mean that the lake is awash in human waste. Meanwhile, it's also being used as a cheap toxic waste dump by the thousands of factories and businesses in the area, as well as by fish-farmers – though one doesn't like to think what is in the fish.

In these cases, we're seeing the consequences of multiple and competing uses being imposed upon the lakes, without any one interest group being able to choose a single dominant use, such as tourism or fisheries, to which all others have to fit in.

MANAGING LAKES: THE SOFT WAY

So are there better ways to relate to lakes? There are probably millions of lakes that are pristine or almost so – 60 per cent of all lakes are in Canada, after all, with close to another 200,000 in Finland. Many of them will have experienced some acid rain and other fall-out from the world's industries, transport, wars and desert dust storms, and some will have had logging operations in their catchments, but most are probably close to a natural condition. A bit of fishing, duck-hunting and summer boating, at worst. But the problems start in less remote areas, where people and their needs come crowding round. Then it becomes vital to consider what a lake actually *is*: an ecosystem that is balanced among inputs and outputs, affected by events throughout its catchment, and capable of producing certain kinds and amounts of beneficial things at a certain rate, which may not be exceeded without eventual loss of those benefits.

A more ecological approach to lake management would recognise that each lake and its associated floodplains, inflowing and outflowing rivers, catchments and ground waters, and their seasonal variation, is a complete system, to be managed as one by and for those who depend upon it, both human and non-human. It would seek to preserve the lake as a lake. Where this aim is endangered by excessive demand, then a decision is needed to use the lake in fewer and less damaging ways than before. If Laguna de Bay, for example, is 'for' sewage, fish-farming, bird-watching and flood control, then all sewage must be treated to become a safe input to a wetland ecosystem that can use it to feed waterfowl and make fish that are safe to eat, and toxic waste dumping will have to stop. Likewise, if Lake Naivasha is 'for' wildlife tourism, flowers and vegetables, then forests and papyrus beds will have to be preserved or restored, organic farming will have to be made universal in the catchment, water will have to be used carefully, and proper sanitation and sewage treatment will have to be installed for farm workers.

The words 'must' and 'have to' are common in those last sentences, but the compulsion comes not from some Stalinist bureaucrat or green fascist, but from the logic and rules of ecology, and the principles of equity and sustainability. The means to the end is important, though, and a new approach would focus on understanding the lake ecosystem as a whole, and then on educational dialogue to build consensus around new ways of relating to it. Revenue sharing, arbitration, new investment, compensation and compulsion may all then have their places in putting into effect the common vision of the lake as a *permanent* ecosystem, rather than as yet another victim of humankind.

7. RIVER WATER

BREAKING THE CHAIN

The dark brown Kama River was getting narrow, half choked by felled trees from both banks. It was a hot, late morning in July 2001, and there was little shade. Surprisingly little. I unrolled the two-year-old vegetation map from the environment ministry in Managua, and checked our position, comparing how far we'd come with what we should have been looking at. According to the map, on the west bank of this river there should have been a band of dense tropical moist forest, several kilometres deep. I could see on the map that this band connected the natural forests of the northern part of south-east Nicaragua to those of the Cerro Wawashan conservation area to the south. Those forests were connected in their turn, so that a squirrel could go a long way north or south from tree to tree, without ever touching the ground. It could find its way into north-east Nicaragua, behind the Mosquito Coast, and then northwards to Honduras, Belize and Guatemala, or it could go south to Costa Rica and eventually the Panama Canal. Such an adventurous squirrel would have been following the 'path of the panther', the *Paseo Pantera* that so many conservationists had been trying to save for decades. Their vision was of a biological corridor, a pathway of connected forests and protected areas between North and South America.

The band of forests on the west bank of the Kama River were the last remaining link between the northern and southern forests of Nicaragua, and therefore ultimately between Mexico and Colombia. But the band wasn't there. From the river all that was visible on the surrounding land was coarse tropical grass, charred tree stumps and logs, a scattering of poor dwellings, and cows. Lots of cows. Small children stared snot-nosed at us from in front of each house. Cows gazed, chewing, from the new paddocks. Horse-flies buzzed. The last link in the chain of forests and rivers between the two Americas had become cattle pasture. It had been colonised by Spanish-speakers from western Nicaragua, despite being deep within an autonomous region that had been set up, at the end of the civil war in the late 1980s, to allow the very different peoples of the eastern half of the country to control their own lands. Clearly, something had gone badly wrong with both conservation and autonomy, and an irreversible disaster had happened. But we were running low on petrol and the boatmen were getting anxious about the return journey to Laguna de Perlas. I told the driver to turn around.

LEARNING FROM RIVERS

Children playing around streams seem instinctively to want to block the flow, by piling up rocks and fallen branches. Then, they often try to cut into the bank to see where the water goes, maybe making a new pond. Many happy afternoons can be spent like this, making the water voles keep their heads down. But by the time they've grown up and graduated as engineers, the games have become more serious, and the new instinct is to pour concrete. As we'll see in this chapter, rivers can be canalised between artificial banks or levees, and they can be excluded from their floodplains. This frees up land for farming and for building houses, and the public applauds. Then, big dams can be created and whole valleys flooded. The resulting lakes can be justified as providing other things that the public wants, such as flood control (even though it doesn't work very well), or irrigation (even though the benefits don't last very long), or generating electricity (even though the lakes fill with silt, and the turbines clog up and wear out).

Those that lose out in these projects can usually be dismissed as unimportant, because they're poor, or haven't been born yet, or because they aren't human and can't vote. Other problems, like falling water tables, eroding estuaries and methane pouring out of the dam lake, are often discovered only much later, when the engineer is working on something else. Like a massive canal to link two river systems, or a project to get rid of islands, shoals and meanders that limit shipping in another river, often creating new benefits for the rich and powerful, and new risks for the less worthy. If anyone complains, well that's what the army's for, isn't it? Meanwhile, other experts are assuring their clients that those particular warehouses, factories or containment dams are so well designed, and so well made, that they're unbreakable and leak-proof under any circumstances likely to happen in the next thousand years.

This is all 'hard', imperial thinking, and gentler minds have been arguing against it for close to 2,500 years. According to former Chinese Vice-Premier Yao Yilin, for instance, quoted in *Damming the Three Gorges* (Earthscan, 1993):

> From the 5th century BC on, Chinese philosophers debated rival theories of river management, mirroring their respective theories of political rule. Taoists believing that rivers should be unconstrained, argued that levees should be low and far apart, allowing the river to seek its own course. Confucians argued for large, high dykes set closely together, tightly controlling the course of the river. This would open up fertile areas along the banks for cultivation, but risked disaster if the levee was breached by floodwaters.

By implication, the softer, more 'Taoist' side of our nature, would be more likely to want to maintain natural meanders and floodplains that sustain livelihoods and ecosystems, to avoid large dams, to restore river flows and fish migration routes, and to return water and floods to river systems. As we'll see, though, the other side is calling the shots in modern China and India, with their giant dam projects and water transfer schemes, as it is among the authorities of the Tennessee Valley Authority, or the US Army Corps of Engineers, or the World Bank. But there are signs of more ecological leanings in Europe, as the constructions used to tame and correct the main rivers are taken apart. This has only happened after repeated and catastrophic flooding, but there are still reasons to hope that we haven't yet forgotten how to go with the flow of a river.

RIVERS AND THEIR SHADOWS

Rivers and streams are the fast part of the global water cycle. Aside from storms that shed vast amounts of rain, huge waves at sea, and migrating herds of wildebeest (each one 65 per cent water) they are also the most energetic. Certainly more so than ice and permafrost, the gradual seep of ground waters, or the puddles of swamps. On the other hand, they run across or through land, which resists them. But the reactive, corrosive, dissolving powers of water mean that land eventually gives way, fragment by fragment. Thus, the Earth's rivers and streams, all 2,000 km^3 of them at any one moment, shape and change the landscapes of the terrestrial world. They shift 40,000 km^3 of water from land to sea each year, carrying 22.5 billion tonnes of the Earth's chemistry and substance. Nine million tonnes of salt are carried down towards Mexico each year by the USA's Colorado River, for example, and 1.4 billion tonnes of silt are delivered annually to the coastal lowlands by China's Yellow River, the muddiest on Earth.

Beneath each river lies its 'shadow' (or hyporheic flow, if you prefer), water that has soaked into the river bed and slowly follows a similar course towards the sea. If the bed and the rocks beneath it are cracked and porous, the volume of shadow water below a river can exceed that of the visible river itself. This trickling mass can also spread more widely, unlimited by the river's banks, feeding ground waters and wells far away. These are further wetted should the river rise above its banks and occupy its floodplain, when the river and its shadow may thoroughly soak a huge area. If floods are common, or the terrain contains waterproof depressions, more permanent wetlands may form: lush, moist ecosystems where the cattle or sitatunga swish through the long grass. Thus the river, an ecosystem itself, creates and merges with other ecosystems, all the way from its source to its eventual entry to the sea or a 'closed' lake, a lake with no outlets.

THE MOUNTAINS OF THE MOON

A river system starts at the precise point where its own catchment is divided from another, where gravity and terrain combine to commit each droplet of rain or dew to one direction or another. Such places are sometimes in landscapes that look so flat that only water can tell which way to run, but more often they're along the crests of mountains. Places like the Continental Divide in the Americas, where rain that falls anywhere between the northern Rockies and the southern Andes is directed either to the Atlantic or the Pacific. Or the boggy, soggy peaks of the Mountains of the Moon, the Rwenzori range in east-central Africa, where water is directed to the rift valley lakes and the White Nile, or to the Congo. But this pre-destined water may not at once become a stream, for it could be held for a while in deep sphagnum moss or peat, and in cold places as ice, before finally being released. If it melts from glacier or ice pack on bare rock, it will tear away downhill. If in a mountain swamp, it may gradually seep away to accumulate in rocky crevices far below the surface, eventually emerging as a spring.

The springs and streams in catchment headwaters seldom have names, since they're far from the homes of people. If they do, it's often because they're felt to be sacred or haunted, where wandering hunters have misunderstood the strange calls of unfamiliar animals. By the time two or three streams have combined, though, far below the misty uplands, the flow is becoming familiar to more settled people, a source of water for cooking and toileting, maybe already a source of fish. From then on, a name is granted, although this often changes as the growing river passes from the knowledge of one people into the lands of another. Lakes provide a good opportunity to re-name rivers, as several may flow in with only one flowing out. Thus, Lake Tana in Ethiopia is fed by the Reb and Gumara rivers, but its outflow is the Blue Nile, which carries the fertile Ethiopian silt that sustained Egyptian agriculture for thousands of years. Name changes also happen at river confluences, for example the river known as the Solimoes from Iquitos in Peru to its junction with the Rio Negro at Manaus in Brazil, is called the Amazon only from Manaus to the sea. And some rivers bear many names, such as the Ganges with 108 of them, from *Puta-tribhuvana* ('The Purifier of the Three Worlds') to the more prosaic *Saphari-puran* ('Full of Fish').

POINT ENDEMICS

People are inveterate namers of things, none more so than those who live traditional lives within ecosystems, and taxonomists. Indigenous peoples need and possess a very detailed understanding of the things they encounter, to help them use and make sense of their environments. Pacific islanders possess complex and practical traditional names for the fish in their waters, such as *chera* for the remora which sticks itself to sharks and turtles, meaning 'clingy woman', or *mam* for the bumphead wrasse with its rich oily flesh, meaning 'fat', or *plutek* for the sound of rampaging sharks swimming in a pack, meaning 'time to get out of the water'.

Sometimes traditional and scientific naming systems resonate in unexpected ways. The Tzeltal people of Chiapas in Mexico, for example, have separate names for different kinds of caterpillar, because they attack different crops at different times of year. A scientific puzzle was resolved in 2004, when DNA testing revealed that the two-barred flasher butterfly was

in fact ten different species that had evolved identical camou-
flage. These ten new butterflies were then traced back to the
caterpillars to which the Tzeltal had already given different
names. Similarly, the Halkomelem Musqueam people of
British Columbia in Canada describe the steelhead and
cut-throat trout as being types of salmon. This was long
thought rather odd, but in 2003 a genetic study revealed that
these 'trout' do in fact belong to the same genus as Pacific
salmon.

Taxonomists, meanwhile, pore over large, small and often
tiny differences among living things, and classify them accord-
ingly. In doing so they make it possible for us to perceive a
little of the true richness of life, and the uniqueness of rivers
and other ecosystems. Without them we'd surely not know
that there are 3,000 fish species living in the Amazon basin.
Or that 77 per cent of the fresh-water fish on the Indonesian
island of Sulawesi live nowhere else in the world. Or that
there are at least four 'point endemic' fish in the wet-zone
highlands of Sri Lanka, which each occur only in a few
hundred metres of stream.

For rivers and streams are ecosystems, but none is the same
as any other. Even if they look identical to us, the salmon
know which one they were hatched in, and return there to
spawn and die. Even at the crude levels of human perception,
rivers are incredibly diverse. A 'typical' river that cascades
from mountains, winds through foothills and then meanders
across coastal plains before reaching the sea, will have unique
features that depend on the structure and chemistry of rocks
and soils in its path. Some have a smooth run of it, cutting
deep into soft rocks and emerging in swampy estuaries, while
others pass through jagged, shelving country, and are rich in
waterfalls.

The Maliau River in north Borneo, for example, is born in
a vast circular basin surrounded by cliffs up to a thousand
metres high. These cliffs were still unknown and uncharted in
1947, when a light aircraft narrowly avoided crashing into
them. Another 34 years were to pass before anyone was able
to find a way into the basin itself, and then they had to use

helicopters. What they found was that the waters of the Maliau River drain a landscape made of inter-bedded layers of sandstone and mudstone, each up to several metres in depth. These are vertically fractured by tectonic forces, and narrow gorges run along the fracture planes to create the main drainage routes. Tributary streams cut down to them across layered rocks, to produce waterfalls at every fracture plane. The result is a spectacular array of falls, the densest in the world, many of them multi-layered. This extraordinary display is driven by up to four metres of rainfall each year, onto the dense rainforests that cloak the basin's walls, and the river escapes from the basin through a narrow gorge cut through its wall.

RIVERS IN BLACK AND WHITE

The water in the Maliau River is the colour of strong tea, not colourless as we might expect after seeing the pristine forests from which it flows. It may be almost free of silt, but is hard to see through nevertheless, has few dissolved ions and is rather acidic. The water picks up these qualities from the piled, waterlogged and tannin-rich vegetation that is slowly decaying in the high-altitude forests around the Maliau basin. Elsewhere in Borneo, there are huge areas of peat in the lowlands too, and black rivers flow from them. And black rivers are common in the Amazon and Orinoco basins, in South America, for the same reasons. The acidity of black water means that ecologically vital ions such as sodium, magnesium, calcium and potassium exist there at a concentration not much greater than they do in rain water. The lack of calcium makes it hard for animals to build shells, so snails and crustaceans are rare in black rivers, as are the fish that would otherwise eat them. One of the largest tributaries of the Amazon, the Rio Negro, takes its name from its black water, while the Solimoes is a silt-laden 'white' and much more fertile river. The swirling mixture of the two waters decorates the Amazon for kilometres downstream from their confluence.

RIVERS IN COLOUR

Other colours than tannin-stain can reveal details of a river's journey. Generally only springs and mountain streams are clear and colourless, while rivers are cloudy, since light falling on them is reflected by silt particles. A similar, sometimes turquoise, effect can occur with tiny bubbles, often where different streams meet, or a river meets the sea. The colour of a river is particularly influenced by the kind of particles it carries, and can be variants of milky-white, green, brown, red or yellow. Glacial melt-waters are often milky-white because they contain rock flour that is ground from the rock surface by the immense weight of the moving glacier. The Yellow River in China is named for the ochre-yellow silt it bears from the Loess Plateau, which is made of ancient wind-blown rock flour. Rivers that are greenish tend to be so because of algae, which show that there's an abundance of nitrogen and phosphorus in the water. And the browns and reds, like the yellows, are due to soils that have eroded into the river, either from the river bed, or by rain falling on bare soil in the catchment. Many tropical lands have bright red soils due to iron that has been weathered and oxidised, and intense rainfall in areas where vegetation has been removed feeds rivers with a characteristic smear of redness, a reliable indicator of logging and farming upstream. A river can pick up many other things too, whatever humans have made available in its catchment or by its banks, and these range from the genuinely scary to the merely grubby.

QUIET FLOWS THE DON

The Novovoronezh Nuclear Power Plant is located in southern Russia by the river Don. Pressurised-water reactors have been operating there since 1964, but those now on line were built in the 1970s and 1980s and a new one is currently being constructed. Cracks were found in the reactor lid in 2000, but repairs were made and the 1970s-era reactors serviced to carry on working until after 2015. Substantial amounts of radioactive caesium-137 have been found in sediments in the River

Don, in forests, and in the Tsymlanskaya reservoir down on the Don–Volga shipping channel, which links the two rivers. All of it can be traced to the Novovoronezh reactors some 500 km upstream. Caesium-137 is an isotope often produced by leaky reactors and weapons tests. It has a half-life of over thirty years, and is easily taken up and distributed within the body. There it produces high-energy electrons while it decays to metastable barium-137, each atom of which sends out gamma radiation for a few seconds before it becomes stable. Various forms of tissue damage and cancer result – silent killers in a quiet river. The point here is that, regardless of the most fastidious safety precautions both in design and operation, *everything eventually leaks*.

RIVERS OF CARRION

While the Russians were studying their cracked reactor lid, the Aurul gold smelting plant at Baia Mare in Romania was spilling a hundred tonnes of cyanide-rich slurry into the Tisza River. Cyanides are compounds that contain the 'cyano group', a carbon atom triple-bonded to a nitrogen atom. This group is extremely toxic because it destroys cytochrome oxidase, a key enzyme in energy metabolism in many kinds of animals and plants. Gold cyanide is soluble in water, so cyanide is often used in purifying that metal, which is why so much had accumulated behind the dam that failed on the banks of the Tisza that day. After the accident, a wave of death passed down the river from the Romanian border into Hungary and onwards to the Danube, killing fish and other wildlife, including dogs that feasted on the dead animals. Officials compared it to another ecological catastrophe in 1986, when fire destroyed a storage building at a Sandoz chemicals factory near Basel, Switzerland. The building held poisonous agricultural chemicals, and water sprayed by firefighters washed about thirty tonnes of pesticides and mercury into the Rhine. This lethal cocktail then flowed onwards through Switzerland, Germany, France and Holland before reaching the North Sea, killing or deforming everything in its wake.

SILVER LININGS

The 1986 disaster, it turned out, was the last gasp not just of millions of fish but of a decade of ineffective efforts to clean up the Rhine. European governments finally got the point, dramatically boosting their goals for the 1987 action plan for ecological rehabilitation of the Rhine. Commitments were made that, by 1992, had cut the amount of cadmium in the river by 34 per cent, mercury and zinc by 47 per cent, the herbicide atrasine (which is implicated in feminising amphibians) by 63 per cent, and polychlorinated biphenyls or PCBs (which are persistent toxic carcinogens) by 77 per cent. The 1987 plan firmly established the idea that the Rhine was a total ecological system, where fish should one day thrive again. By 2006, €60 billion had been spent on water purification, at least sixty species of fish had returned to the river, and the Rhine had become a World Heritage Site. The remaining problems largely involve agrochemicals leaking from ground water. To solve these would require much broader reforms to agriculture throughout the Rhine catchment, including the wholesale introduction of organic farming.

DAMMING TO ABSTRACTION

Before we discovered fossil fuels, we had water wheels. The early stages of the Industrial Revolution in Britain were driven by moving water, with tens of thousands of wheels being turned by streams channelled through specially built mechanisms. The momentum of flowing water was systematically harvested to grind, pound, stamp, hammer, spin and screw manufactured value out of raw materials across the land. Then as the midpoint of the nineteenth century approached, the sheer density of energy stored in coal, combined with new generations of steam engines to exploit it, caused a steady abandonment of water power for most industrial uses.

It took electricity to renew interest in water as a source of industrial-scale energy, and the first hydroelectric dam went up in the 1880s. While fossil fuels burned within machines could do far more work than water power obtained locally, electricity could be created in one place and then wired to thousands of machines, where civilisation could be mass-produced. If a cheap dam could harvest energy from the height of a dam lake, and change it into electrical energy by blasting water through rotating turbines, then the industrialists and politicians were interested.

Once the technology had matured, hydroelectric dams on rivers began to spread. There are now tens of thousands of them, together generating close to a fifth of the world's electricity. Because they're limited by the capacity of people to build dam walls that are high and strong enough, even the largest dam lakes contain not much water compared with many natural lakes. Among the biggest, the Kariba dam lake in Zambia and Zimbabwe holds 180 km^3, the Aswan in Egypt and Sudan holds 157 km^3, and in the USA the Glen Canyon and Hoover dam lakes hold 63 km^3 and 35 km^3 respectively (when full, which they often aren't these days). Some dam lakes are quite huge though, by the standards of the lands in which they are set. The largest, the Akosombo, flooded $8,500 \text{ km}^2$ or nearly 4 per cent of Ghana's land area, while Mozambique's Cahora Bassa dam lake flooded an area of $2,700 \text{ km}^2$, and Brazil's Tucuruí flooded $2,430 \text{ km}^2$.

I said before that large bodies of fresh water are precious, with all sorts of uses and reasons to be valued. But with artificial lakes, the values are all economic, and the uses are all about control. By building a dam, planners, engineers and political leaders are deliberately sacrificing an area of land, usually inhabited by people and always occupied by ecosystems and wildlife, in favour of a lake. That lake has the potential to do at least three things that a decision-maker's constituents might value. It can generate electricity, it can prevent floods, and it can yield a flow of water for irrigation.

The problem is that it can't do all three at the same time. For electricity, the dam lake needs to be as full as possible, so maximum energy is available to spin the turbines. But if the lake is full, it'll overflow if flood waters run into it and the flood will continue downstream. And irrigation water taken out of the lake, extraction of which will need to peak in dry months when the lake isn't full enough anyway, is in direct competition with the electricity generators. To meet the promises that are made to voters and financiers in order to get the dam built in the first place, the lake would need to be full all the time, while also being half empty in the wet season, and more than half full in the dry season. Dam management is an art of compromise.

WEIGHT OF EVIDENCE

Large dams have other problems too. At a billion tonnes per cubic kilometre their lakes are heavy, which is a new weight in a new place for the Earth's crust to bear. It can cause subsidence, earthquakes and landslides, all of which can cause a dam to fail, crack, leak or overflow. We were taught some of this by the Vaiont dam disaster in Italy, which killed about 2,500 people one night in October 1963. This dam, 100 km north of Venice, has an enormously high wall, at 262 metres. Smallish landslides occurred while it was first being filled, but it was decided that the geology could be stabilised by filling and emptying the dam a few times. Stresses from this process, combined with the effect of heavy rains, caused 260 million cubic metres of trees, earth and rock to slide suddenly into the reservoir. A wave of water was pushed up the opposite bank and destroyed the village of Casso, 260 metres above lake level, before over-topping the dam by up to 245 metres. About 50 million cubic metres of water then fell more than half a kilometre, onto the villages of Longarone, Pirago, Villanova, Rivalta and Faè, destroying them utterly. The dam was largely undamaged by all this and is still standing today, with its by-pass tunnel being used to generate electricity.

DEADLY INDIRECTIONS

Hydroelectricity is often seen as a clean source of power because it doesn't produce carbon dioxide. But in fact, making every tonne of cement used in dam building creates about a tonne of CO_2, and more is emitted in making steel and other components, and by the vehicles used in construction, while the decaying vegetation in a dam lake can continue producing methane, a very potent greenhouse gas, for decades. Dam lakes also act as silt traps, since the mud carried along by fast, inflowing rivers settles to the lake bed when the water slows down. This can turn a dam lake into a useless swamp with remarkable speed, especially if logging and farming is going on in the lake's catchment. Few dam managers control land use throughout the entire drainage basin that feeds their lakes, and big, new lakes can themselves provide new ways to reach remote areas by boat. This can allow farmers and even loggers into places that they'd never have reached before, with unexpected impacts on forest cover.

It can also let hunters into new areas, with impacts on those wildlife populations that have already survived the drowning of a large part of their habitats. Because rivers are at the bottoms of landscapes, they are at relatively low altitude compared with the hills and mountains around them. Especially in tropical forests, lower areas tend to be more productive and to have a far greater number of wild species than the colder and drier uplands. So, as a dam lake fills, it will submerge the richest and most biodiverse ecosystems of the landscape. While there have been attempts to rescue animals swimming in expanding dam lakes, and to release them on higher land, in fact the new location will already be fully occupied by that species, or else will be unsuitable habitat for it, so the enterprise is usually pointless. Meanwhile, dams act as giant barriers to the dispersion of fish and other animals up and down rivers, breaking up their populations and preventing migrations and spawning runs.

Large dams often have dire effects on people too. India's many large dams, for example, have displaced somewhere between 16 and 38 million people in total, 52,000 here (the Ukai dam in Gujarat), and 90,000 there (the Pong and Bakhra dams in Himachal Pradesh). The numbers add up, and each represents individuals, families or communities uprooted to an entirely new site, whose way of life has been completely changed, seldom for the better. Meanwhile, once the first social engineering has been accomplished, there are impacts on downstream communities as their river is reduced while the dam is filling, and then flows erratically as water is released through the turbines in response to demand for electricity in cities and factories far away.

Some dam managers try to flush out sediments by suddenly releasing water, as is planned for the Three Gorges dam in China, or they may have to lower the lake's level quickly if there's heavy rain on the catchment and the risk of an overflow. The people living downstream are seldom warned and can be caught in the ensuing floods, as has happened in places as far afield as India (the Hirakud dam), China (the Banqiao dam), and Nigeria (the Kainji dam). It almost goes without saying that large dams can be a tempting target during war or for terrorist attacks, since disabling one can reduce a country's power supply, and breaking one open can do immense damage downstream as well.

There are now more than 45,000 large dams in the world, which take around 3,800 km^3 of fresh water annually from the world's rivers, lakes and aquifers. Three-quarters of them are in just five countries: China, with about 22,000, the USA with 6,400, India with 4,000, Japan with 1,200 and Spain with 1,000. The total cost of building them is thought to be around two trillion dollars, and in developing countries much of the finance came in the form of loans, from the World Bank and other organisations such as the Asian Development Bank or Inter-American Development Bank, or else was subsidised in other ways by governments.

Construction peaked in 1970–75, when nearly 5,000 large dams were built worldwide, and the rate of construction has since eased off greatly in the USA and Europe. But the era of big dam building isn't over yet, despite all the problems that have arisen. The colossal Three Gorges dam on the Yangtze River in China is nearing completion and will eventually hold about 39.3 km^3 of water, while construction of the Bakun dam on the Balui River in Sarawak, which will flood an area of 700 km^2 of Borneo's rain forests, has now been reactivated after being on and off the shelf since the 1980s.

A NEW DAM FOR AFRICA

Uganda is an attractive place for dam builders, especially the Victoria Nile, which falls steeply as it drains from Lake Victoria at Jinja, on its way north and west to Lake Albert. This section of the river already has a hydroelectric dam complex, the Nalubaale and Kiira power stations, which incorporate the Owen Falls dam, built in 1954, and an extension canal built in the 1990s. Further proposals were made in the late 1990s to build a new dam eight kilometres north of Nalubaale and Kiira. The aim was to re-use the water released from the upstream dams to generate an additional 250 megawatts of electricity. A 22-metre wall would hold back a new lake, 4 km² in area, that would drown the Bujagali falls, a spectacular series of cascades.

Agreement on a World Bank loan for the project was reached in 2001, but was later delayed by a corruption scandal and then cancelled when an investor pulled out. Negotiators then pieced together a fresh agreement on a complex financing package for the dam and its electricity distribution system. This was to involve the government of Uganda, European donor agencies, private investors, the African Development Bank and the Washington-based World Bank Group. This latter includes the Bank itself, which lends to governments at normal interest rates, the International Development Association or IDA, which makes soft loans to poor countries, the International Finance Corporation or IFC, which lends to the private sector, and the Multilateral Investment Guarantee Agency or MIGA, which protects private investments. The investment package, worth about US$800 million in total, was finally activated in April 2007, when the Board of Directors of the World Bank Group approved US$130 million in IFC loans, and over US$230 million in guarantees from the IDA and MIGA.

During this process, the Bujagali project was challenged by those who held that existing dams on the Nile had contributed to the shallowing of Lake Victoria, now at a record low, by taking too much water for making electricity. Critics also made the point that 95 per cent of Uganda's population has no access to electricity, and most could not afford it even if they were connected to the national grid. Hence, they argued, the Bujagali dam will not bring power to the rural poor, and it would be better to use other forms of electrification, including improving the existing electricity delivery system, and decentralised solar, microhydro and geothermal power. In Kenya, they pointed out, more households got their electricity from the sun than from the national grid.

The original 2001 agreement between the World Bank and the government of Uganda had made a link between financing the Bujagali project and protecting the Mabira forest reserve, which forms part of the catchment area for the dam. The idea was that this would conserve biodiversity in compensation for that destroyed by the dam and its lake. Mabira is an important place for conservationists, being home to at least 300 bird species and 60 mammals, including an endemic mangabey and other rare primates, as well as the hero shrew, which can survive a man standing on its back. It's also a source of livelihood for more than a million people who depend on it for water, firewood, honey, mushrooms and materials for making baskets and mats. As the final loan negotiations proceeded, however, conservationists began sending each other worried emails about whether or not the protection of Mabira would be maintained as a condition of the loan, which seemed not to be the case although the bank later insisted that it was. This now seems to be in doubt.

Although this condition seemed to have lapsed, the Bank later clarified that, as far as it was concerned, Mabira was still safe. The issue had been blurred by a Ugandan government policy shift during 2006, whereby several forest reserves were to be given over to agricultural enterprises, including a quarter of Mabira (71 km²), which was to be used for growing sugarcane to meet international demand for biofuels. This new policy was reiterated in July 2007 by Uganda's president Yoweri Museveni, so there remains a question mark over the future of this small but vital forest.

THE RIVERS OF DIPLOMACY

Visible rivers, shadow rivers and floodplains are all altered by dams, and wells and grazing lands may dry up far from the river's course. If the dam is just for water storage, there may be no downstream flow at all, especially in dry areas where the dam lake is emptied by evaporation and irrigation. If it is for irrigation, a large area of dry land may be farmed for the first time, but as we saw in Chapter 6, irrigation may be a pestilence in disguise. It creates multiplying interest groups whose livelihoods and prosperity depend on maintaining the flow of water, and if the flow is compromised by over-use or drought, competition for water supplies can be harsh.

The case of the Colorado River in the USA's south-west is a classic example of powerful interest groups (including seven US states and Mexico, and many of the large plantations and cities they contain) all dividing and sub-dividing (not often amicably) a river's flow until there's nothing left but salty dust. Meanwhile the irrigation water itself does its mischief by waterlogging and salinising soils to a point when as much water is used to wash off the salt before planting as is used for irrigating the crop (as in Uzbekistan), or when the fields of a nation's breadbasket may soon have to be abandoned (as in Pakistan's Punjab).

Capturing a river in a dam lake, and using its water locally, means that the country possessing the dam may automatically deprive another country of water. This may be done with aggressive intent, or in the spirit of national assertiveness, or just because a government's first duty is to its own electorate rather than to those of its neighbours. An emerging classic case is Turkey's use of the waters of the Tigris and Euphrates rivers, which both originate in its land. Since 1990, when the Atatürk dam and river diversion scheme was completed, another 22 large dams have been built as part of the South-East Anatolia Project on these rivers, while others, such as the Ilisu dam on the Tigris, are under construction. The net effect will be to irrigate 1.7 million hectares of Turkey and provide the country with extra electricity, but the downstream countries of Syria and Iraq are losing out, inevitably and disastrously. As Douglas Jehl wrote in the introduction to *Whose Water Is It?*, 'Under the hot summer sun, a Syrian farmer who still hopes for the day the Euphrates will reach his fields tells a visitor that he used to pray to God for help, but now he fears that is not enough. "It's not God who has our water," he says. "It's the Turks." '

The potential for friction over shared rivers is immense, since 254 major river basins are shared by two or more countries worldwide. Some, such as the Danube, Congo, Niger, Nile, Rhine, Zambezi, Amazon, Ganges-Brahmaputra, Jordan and Mekong, are shared by six countries or more. Events inside each country's share of the catchment change the river downstream. They can include deforestation and erosion, irrigation and agrochemical use, the discharge of toxic effluents, the escape of leachates from garbage dumps, and the release of untreated sewage, as well as the building of dams and canals and the diversion of water to cities. The scope for dispute is so great that 'water diplomacy' is an important job of foreign ministries in many countries.

Some diplomatic processes, though, have been able to turn potential disputes into instances of international co-operation. The Mekong River Commission, for example, was established in 1995 and has encouraged dialogue and joint problem-solving among four of the Mekong River countries – Cambodia, Laos, Thailand and Vietnam. In 1996, discussions were extended to include China and Burma as well. China refused to join the commission, however, amid tensions with its longstanding rival Vietnam, which has argued that China's construction of eight hydroelectric dams on the Mekong by 2020 will devastate its downstream ecology. The Mekong already generates power in its upper reaches and tributaries in China, Thailand and Laos, and sustains the rice production and fishery systems that support more than 60 million people in the lower parts of its basin. Since the upstream dams limit water flow, or send irregular pulses downstream, and block fish spawning migrations, the whole system is intimately linked and international co-operation is vital, although effective agreements remain, as ever, elusive.

CONTROLLING RIVERS: GUIDED WATER

Large dams are far from being the only massive constructions that people have made to try to make sure that water goes where it's supposed to, and in the right amounts, for human use. We have already encountered the canal between the Don and Volga rivers in Russia, which was designed to shorten shipping time between the two basins but also allowed radioactive material to pass between them. Other such links between rivers, or between rivers and lakes and cities, are common worldwide. Aqueducts have been built since ancient times, most enthusiastically by the Romans to supply cities throughout their empire, but also by earlier peoples such as the Assyrians and Egyptians, and others in India, who invented most of the key techniques for their design and construction. These techniques were lost with the fall of the Roman Empire, and aqueduct building only started again on a large scale in the nineteenth century, to irrigate the Industrial Revolution. Modern aqueducts include many in the USA, such as the Catskill aqueduct, which waters New York City, and the vast Colorado River aqueduct.

As the distribution of fresh water on land increasingly ceases to match the location of cities and industries, because of local over-use or climate change, projects to move water over long distances are becoming increasingly common, and bigger. In India, various parts of the River Interlinking Project are under construction, with the aim of distributing about 47 km^3 of water each year from the Ganges and Brahmaputra to seventeen rivers in the south of the country. On a similar scale is China's South-to-North scheme, which will move about 45 km^3 of water annually from the Yangtze River system in the south to the thirsty cities of the north.

The Chinese scheme has three parts, the first two of which are under construction and the third is being planned. In the first, water will flow from the Danjiangkou reservoir by canal to the North China Plain. In the second, it will pass from the mouth of the Yangtze River north to Shandong province through a repaired Grand Canal, parts of which were first built up to 2,500 years ago. In the final part, water will flow from the headwaters of the Yangtze in Tibet, through tunnels blasted through mountain rock into the headwaters of the Yellow River. Both the Indian and the Chinese projects will move as much water each year as the Karakum canal which devastated the Aral Sea, and may well have comparable impacts on water, people and ecosystems.

RIVERS IN STRAITJACKETS

As a youngster I went on a school journey to the USA, and visited the headquarters of the Tennessee Valley Authority, where I was impressed by a vast concrete model of the entire river system, complete with little obstructions on its bed to simulate turbulence and water speed in different places. The TVA was set up in 1933 to correct the river's unruly flooding behaviour, as well as to generate electricity. It inherited one dam on the Tennessee River and from 1936 to 1942 built another seven on it and twenty more on its tributaries, with a final major dam being completed in 1967. This barrage of dams is claimed to provide flood protection to more than 2.4 million hectares of land, while reducing the frequency of flooding on another 1.6 million, but the evidence isn't wholly convincing. The Tennessee River flooded seriously enough to be recorded in local histories on average once every three years and ten months between 1808 and 1932, just before the TVA was created, and once every *two* years and ten months between 1933 and 2003. Thus the frequency of flooding seems at best unaffected by the dam-building programme in the Tennessee basin in the 1930s and 1940s.

But the Tennessee River is only a small part of the great Mississippi system, which is fed by other big rivers and drains 41 per cent of the continental USA. Dams, locks, canals and artificial banks were built by the US Army Corps of Engineers throughout the system during the twentieth century, with the aim of enhancing navigation and controlling floods. In the 1950s, government scientists decided that the lower Mississippi was trying to join the Atchafalaya River as an easier path to the sea. This would have isolated New Orleans from the main river flow. The Old River Control Structure put a stop to this escape, but the force of the Mississippi denied its way was so strong that it had to be supplemented by another flow control station, built in 1986. As things turned out, New Orleans might have been better off on a side channel of the Mississippi, rather than right on the main river, when Hurricane Katrina came calling in 2005.

Meanwhile, the innumerable flood control works that had been built further upstream were having a profound effect on the river's behaviour. To stop water getting into its floodplain, artificial levees up to 15 metres high had been constructed on at least 10,000 km of the Mississippi's banks and those of its tributaries. The river had also been straightened by cutting through meanders, and for 1,750 km it flowed through artificial channels. In 1992, a federal government inter-agency study of floodplain management concluded that sixty years of building flood control structures in the Mississippi basin hadn't had any real effect in reducing deaths and property damage. The very next summer, after most of the river's catchment had received up to 200 per cent more rain than normal, water sped downstream along the straightened and restricted channel, and hit the city of St Louis where the Mississippi and Missouri rivers meet. The river brushed aside the levees that hemmed it in, and 487 counties in Illinois, Iowa, Kansas, Minnesota, Missouri, North and South Dakota, Nebraska and Wisconsin became flood disaster areas in a matter of hours.

CORRECTING THE RHINE

Similar tales are carried in the waters of the main rivers of Europe, such as the Danube and Elbe, and especially the Rhine with its 6 dams, 68 large cities and only 7 per cent forest cover in its catchment. This river was most unsatisfactory to engineers and has a long history of being 'corrected', especially in Germany. Here, in the late nineteenth century, the largest civil engineering project ever attempted shortened the upper Rhine by a quarter, hard-banked much of it, and removed thousands of islands and peninsulas. This created a shorter, straighter, faster and more navigable river, trapped between artificial banks, but at the cost, when the salmon stopped coming, of an entire society that depended on river fishing. And this taming of the upper Rhine continued for another century, by which time some 85 per cent of the floodplain had been lost, and the wetland corridor maintained by the river and its floods had been narrowed to a belt less than 150 metres wide. Ecosystems along the river changed fundamentally, mainly from forest to farmland and built-up areas.

The behaviour of the upper Rhine also changed, in ways that the engineers didn't expect. By the end of the first phase of river correction, the river's speed had increased so much that its bed was being scoured out, in some places down to bedrock, and both the river and water tables were falling in many areas. Then again, flood peaks travelling down from the upper Rhine were much faster than they had been, and began to slam into the middle and lower parts of the river, causing serious flooding in places that had never been flood-prone before. From the 1940s to the 1980s, further corrections to the main river and its tributaries, especially canalisation to improve navigability, had more than doubled the speed of flood peaks. It had also increased the chance of multiple peaks from the Rhine and its tributaries such as the Rench, Kinzig, Murg, Ill, Moder, Sauer and Neckar, reaching the towns of the middle and lower river simultaneously.

The result was a succession of devastating floods in the lower reaches of the Rhine, in Germany, France, Belgium and the Netherlands, in 1983, 1988, 1993, 1994, 1995, 1998, 1999 and 2002. By the late 1990s the Europeans had got the point, and were trying hard to solve the problem, driven by the massive economic costs of the flooding. An important part of their strategy was to undo many of the earlier engineering works around the Rhine and other rivers, thus 're-activating' natural flood storage by restoring the rivers' access to their floodplains. This, combined with improved flood warnings and a better focusing of local flood protection structures, had improved matters by 2006. In April of that year, for example, Saxony State Premier Georg Milbradt observed that 'we are not experiencing a catastrophe as we did in August 2002. The damages are not comparable, and we are better prepared. Right now we are having a winter flood that is stronger than normal.' The work of 150 years cannot be corrected overnight, or without expensive changes to settlement and land use around rivers, but it seems that progress can be made.

LIVING ON FLOODPLAINS

Calls for a similar approach to river flooding in England are becoming louder too, but with little obvious effect so far. In June and July 2007, flooding devastated vast areas of England, first in Yorkshire and the East Midlands, and then across the west, centre and south of the country. Insurance claims for damaged homes and businesses were likely to exceed £3 billion (US$6 billion), but the total economic cost was estimated to be at least twice that, and the long-term impact on property values may prove to be far greater still. The immediate cause was a southerly displacement of the Atlantic jet stream, which caused sustained, intense rainfall over England and yielded the wettest May to July ever recorded. In the bigger picture, though, this is consistent with the expectation that global warming would increase storminess and rainfall intensity in maritime locations, like the British Isles, as the warmer sea provides extra heat and water vapour to fuel storms and downpours. This has long been expected, but climate science has done little to influence land use planning in England, which for decades has encouraged reduction of the landscape's capacity to absorb water, by covering huge areas in tarmac and concrete, as well as the construction of large numbers of buildings on floodplains. These efforts simultaneously reduced the rate at which rain would have to fall in order to create a flood, and ensured that any flood would cause maximum damage to life and property.

Meanwhile, as global warming continued, the atmosphere became warmer and wetter, and therefore more turbulent and storm-prone. On an imaginary graph, the lines representing a vulnerability to floods and a tendency to storminess intersected for England in 2007. Since then, insurance companies have been arguing with government over access to information on flood defences, and have started to threaten to make vulnerable locations uninsurable. So it is beginning to make sense to stop building in floodplains and to relocate vulnerable structures, while also restoring the country's ability to absorb water as much as possible. In short, a much more cautious and ecological approach to life in and around floodplains and catchments would be a wise response to harsher and less predictable weather.

POVERTY AND RAIN

Unjust Waters is a 2006 report by the charity ActionAid, which focuses on the experience of flooding among people living in African cities. It makes sobering reading, as all the factors we've looked at come together, including ecological damage to catchments, wholesale construction in floodplains, poor design and maintenance of drainage systems, and, hanging over everything, climate change. In August 2006, in Ethiopia, the overflowing Dechatu River hit Dire Dawa town at night, drowning 129 people and wiping out 220 homes. Flooding is a major problem in all informal settlements in Nairobi, Kenya, and in the Maili Saba slum, for example, which is part of Dandora and next to the Nairobi River, flooding is a normal occurrence. The houses of the poor are weakly built and prone to be swept away, while new settlers have built many additional, vulnerable houses close to rivers.

But even where flooding is normal, it may not be predictable any more. As Mrs Fatu Turay of the Kroo Bay community in Freetown, Sierra Leone, put it: 'The pattern of floods has been changing every year. The worst flood this year was in June. The climate is changing. The rains have been coming more than before and the weather has been getting hotter.' In this community, much of the problem comes from changes in land use on the hills outside Freetown, partly due to pressure on the land by people displaced by the civil war of 1991–2002, made worse by urban development that encourages water to run off quickly into low-lying areas like Kroo Bay.

In Kampala, Uganda, unregulated building of slum areas like Kalerwe, Katanga, Kivulu and Bwaise has stopped water soaking into the ground and increased runoff by 600 per cent. Since the 1980s, flooding in these places has become much more frequent, and accompanies every small downpour, aggravated by drainage channels becoming blocked by silt and debris. Climate change is also a factor, according to local Kampala people, who say that flooding used to occur in predictable cycles in the two main rainy seasons of April–May and October–November, but is now much less predictable, as well as more frequent and more severe.

In Accra, Ghana, women in the Alajo area tell a similar story, that patterns of rain and flooding have become unpredictable since the 1980s. They note that it used to rain heavily in June and July but, since 2000, the heavy rains sometimes start earlier than June and in other years after July. And for these people, flooding is an uncomfortable, even dangerous, reality, forcing them to leave home whenever the clouds gather, to spend nights clinging sleeplessly to the tops of wardrobes, and to suspend the activities that, even at the best of times, earn them only a pittance a day. The immediate impact is the loss of livelihood support for subsistence needs, and for children's education and health bills. As a resident of Alajo put it, 'When the rain and the floods come, women and children suffer. You can be locked up for up to two days with the flood. Sometimes we take our children out from the room to the rooftop. Then people bring boats to evacuate people.' Those whom ActionAid interviewed added that their complaints to government authorities brought no results.

The *Unjust Waters* study concluded that:

Climate change will increase the vulnerability of the urban poor throughout Africa. Already the urban poor are forced to live in hazardous places. Many build their homes and grow their food on river flood plains in towns and cities. Others construct their shelters on steep, unstable hillsides, or along the foreshore on former mangrove swamps or tidal flats. Whether already vulnerable to destructive floods, damaging landslides or storm surges, climate change is making the situation of the urban poor worse.

As always, the overall crisis is made up of thousands of local crises, where individuals, families and communities struggle daily with the behaviour of local ecosystems and rivers. Although it's easy to feel overwhelmed by such threats, and excluded from decision-making, slum-dwellers in Nairobi proved to have many ideas about what they could do about it. These included forming residents' associations to improve their own welfare and their response to emergencies, and joining with others to plant trees along riverbanks, and to dig canals, trenches and drainage next to their houses. They also imagined joining forces to make landlords build better and more flood-proof houses and business premises, further from the river. As ever, human solidarity is a key part of all solutions.

8. GROUND WATER

THE LOUDEST NOISE EVER HEARD

Our blue and white, green and brown planet shimmers in space. Imagine swooping down almost to sea level in the Indian Ocean at the equator, and bearing a little south of east for a thousand kilometres. We slow as we see the hilly islands of Sumatra and Java ahead of us, with the Sunda Strait between. Over there is a disturbance in the white cloud, a dark smudge with its source just over the horizon. Rising a little, we creep closer until we can see what's happening. An island mountain is belching smoke and steam from a vent close to its base, seemingly slowly at this distance, but closer the violence of it all becomes clearer. The island's name is Krakatau, and it's early in the morning of 27 August 1883.

As we watch, the mountain seems to pause and tremble. We can almost hear the roar as millions of tonnes of sea water flood the fractured roots of the island. Then the mountain explodes with unimaginable loudness in an expanding steam-driven shockwave of shattered rock that tears past and through us. The darkness closes over Krakatau, but already 30-metre-high waves are breaking over the nearest islets on their way to the main islands. Buffeted by turbulence we wait further. Seventy minutes later the mountain explodes again, and then again, and then, over four hours since the first blast, comes the greatest sound ever heard by humans, and Krakatau is gone at last, the steaming sea crashing back and forth across its remnants.

Sunsets around the world will be spectacular for years, until the dust of what was Krakatau has settled out of the atmosphere. But even before the tsunamis have finished their work on the coastal villages and towns of Sumatra and Java, we rise back into near-Earth orbit over Indonesia, glancing left and right as we evade the volcano cloud. Through the haze we see volcanoes marching north-west along the spine of Sumatra, and east-south-east along Java. We can see Krakatau in context as one point in a great arc of vulcanism. We wonder about the titanic confrontation we've just seen between ocean water and Earth fire, with no doubt that the water entering the hot interior of the mountain must have vaporised and expanded to blow it apart. But what role might water play in the longer term, in this unstable landscape?

SUCH STUFF AS ROCKS ARE MADE OF

To understand, we need to look both at rocks and at how the surface levels of our planet are organised. First, rocks often contain *hydrous minerals*, which incorporate water within their molecules. These are formed when surface waters seep underground, dissolve and react with various solids, and are then heated, pressurised, cooled and dried over time. Familiar examples are rock salt, gypsum and opal, but the many others include anhydrite, torbernite, olivine, serpentine and clay minerals such as chlorite and kaolin. Gypsum contains two molecules of water for every bonded molecule of calcium-sulphur-oxygen, while torbernite has up to twelve for each complex molecule of copper-uranium-phosphorus-oxygen. The point is that these rocks are largely water, and they stay largely water even at considerable pressures and temperatures. In more extreme conditions, though, up to about 320°C, other kinds of hydrous minerals are formed as crystals, including the marbles, micas and quartzites such as amethyst.

So, the upper crust of the Earth contains vast amounts of water, bound into minerals and crystals. But the upper crust isn't static or permanent; it moves and it sinks. It does both because it's made of huge plates, scores of kilometres thick, cool on top and grading towards moltenness below, that float on a hot, fluid interior. This interior has convection currents, rotation currents as the Earth spins and tidal currents as well. The floating plates on the surface therefore move, and jostle against each other in slow motion. Vast masses are involved so the moving plates have huge momentum. When two of them meet, they either crumple into mountain ranges, such as the Himalayas, or one must force the edge of the other down into the fire. This all happens deep below the surface and is accompanied by much grinding, sticking and sudden jarring movements, which we above experience as earthquakes.

When the surface crust of a plate, with its freight of hydrous minerals, is squeezed downwards, pressure and temperature start to rise dramatically as the hard edge of the sinking plate drives into the hot, soft underbelly of the dominant plate, and the minerals begin to lose their water. This is boiled off along the slope of the sinking plate, and has a dramatic effect on the hot, almost-liquid rock above it. The water pushes the almost-liquid rock over the edge, to become fully liquid; in other words, it lowers its melting point. Having melted, this band of water-affected liquid rock is less dense than the solid rock above it, so it tries to float. Since the clashing of the plates has created many cracks in the deep crust, the molten rock finds ways to squeeze upwards. As it rises, it expands further as the pressure declines, and it forces the cracks wider and wider until it eventually reaches the surface. This is where the fire of Krakatau came from, which the sea tried to extinguish with such cataclysmic results. It's also why there is a line of volcanoes right along Sumatra and Java, for this is the edge where one plate is being forced beneath another. And all these volcanoes, whether on land or sea, are driven by water boiling out of hydrous minerals far underground, as the Earth's crust is recycled.

Much of the water that's driven out of minerals, crystals and the remains of organisms at lesser depths returns through volcanoes to the atmosphere, where it re-enters the realm of life. Once there, it unites with atmospheric water vapour to fall later as rain. We've seen how this rain feeds springs, streams, rivers, shadow rivers, glaciers, lakes and all the world's ecosystems as it passes back to the sea or back to the air. But some rainwater soaks deeper into the Earth and loses contact with events on the surface. This is ground water, the ultimate source of hydrous minerals but also one of the great stores of water for the biosphere and its denizens, including people. For in the deep soils and rocky beds of the world are estimated to be close to 12 million cubic kilometres of liquid water, almost all the liquid fresh water on Earth and twice as much water as there is in the ice of all the planet's glaciers and polar regions combined.

AQUIFERS

Just as the name of the tree genus *Mangifera* means 'mango bearer', so *aquifer* means 'water bearer'. Aquifers are aptly named, for they are underground layers of water-bearing permeable rock, gravel or sand, the upper limit of the saturated zone being the water table. They are fed by water that soaks into and down through the soil, and continues to penetrate until it's brought up short by a layer of waterproof rock. Then, if the barrier slopes away downhill, the water will follow it, maybe being released later as a spring if the rock eventually reaches the side of a valley. Or perhaps the bed of a lake or a sea shore, to produce an underwater spring. Or maybe the layer of water will flow into the shadow of a river, uniting with it and contributing to the workings of the river and its floodplain. Where flowing water will go is subject only to gravity, the permeability of the rock, and the capillary action caused by its own surface tension sucking on surfaces in narrow spaces.

However, if the waterproof rock is bowed, so the water cannot flow away, then it will puddle underground, forming a lake of saturated stony material, where every crevice and pore is full of water. In reaching such a place, the water may have approached across the surface of one layer of waterproof rock and become trapped beneath another. Squeezed by the weight of overhead rock, the deep water will become pressurised. This will also happen if the various layers of rock among which water is dispersed are deformed by the slow pressure of plate against plate in the Earth's crust. This is all agonisingly slow, but the net result in the real Earth is a jumble of streaked, layered and puddled water at all sorts of depths and all sorts of pressures and temperatures.

EXPORTING AQUIFERS

Many aquifers are potentially renewable as water resources, and could produce water at a certain rate for ever, or at least until climate change stops their recharge, the entry of new water. This is more or less the rate that people once used them, before we discovered how to burn fossil sunlight, such as coal and oil, to extract fossil water directly from underground, using mechanical pumps. These can take water from an aquifer very much faster than it can be recharged naturally. The danger usually comes when large numbers of people use powered drills to send many boreholes down into a single aquifer, and powered pumps to suck water out of it. These pieces of equipment used to be rare and expensive, but since the 1980s tens of millions have been made and distributed around the world, with prices reduced by mass production and the equipment often subsidised or given away by donor agencies or governments in order to encourage farming. The result has been a worldwide 'ground water revolution', in which millions have become dependent on water drilled and pumped from underground.

Technology is only part of the story, though, since insatiable markets are also needed for over-exploitation to happen. This is where 'virtual water' comes in – the amount of water used to make things. We saw before that it takes about 2.8 tonnes of water to make a single cotton T-shirt. Well, on average, to make a kilo of wheat takes about 1 tonne of water, of sugar 3 tonnes, of milk up to 4 tonnes, of rice up to 5 tonnes, of coffee 20 tonnes, and of beef 24 tonnes. All of this adds up to 1,000 km^3 or a trillion tonnes of virtual water being traded every year. This demand is more than enough to drain many aquifers when it's combined with open access and cheap technology.

Water can also be exported more directly, through a form of international parasitism. The UN's *Human Development Report* of 2006 shows this happening. It lists countries that are using water faster than their own total renewable water availability, and compares this with the amount of water they receive from outside their own borders. Countries using 100 per cent or more of their own water, but receiving none from outside, include Malta (using 100 per cent), Barbados (113 per cent), Oman (138 per cent), Yemen (162 per cent), Libya (711 per cent), Saudi Arabia (722 per cent) and the United Arab Emirates (1,553 per cent). The first three are using moderate amounts of ground water, the others exceptional quantities from their own aquifers. Some other countries use a great deal of water, but this is mostly taken, one way or another, from other countries. These include Turkmenistan (94 per cent imported, diverted from the Amu Dar'ya River and hence the Aral Sea), Uzbekistan (68 per cent imported, likewise), Egypt (97 per cent imported by the Nile, from Sudan and beyond), Israel (55 per cent imported, from occupied territories, catchments and aquifers) and Bahrain (97 per cent imported, from the Eastern Arabian aquifer in Saudi Arabia).

BOTTLING AQUIFERS

Sales of bottled drinking water have been growing at a rate of up to 10 per cent annually since the mid-1990s, and are now approaching 0.2 km^3 each year, or 200 billion litres with a retail value of about US$100 billion. Bottling corporations have focused on acquiring water sources, usually ground waters so the product can be sold as 'spring' water, and in marketing it as being 'purer' than tap water. In 2006, about 32 per cent of all bottled water was sold in Europe, 30 per cent in North America, and 26 per cent in Asia. The countries that consumed most in 2004 were the USA (26 billion litres, which rose to 31 billion in 2006), Mexico (18 billion litres), China and Brazil (12 billion each), Italy (11 billion), Germany (10 billion), France (9 billion), Indonesia (7 billion) and Spain (6 billion). In that year, the number of litres used per person was highest in Italy (184), followed by Mexico (169), the United Arab Emirates (164), Belgium and France (145 each) and Spain (137). In 1999–2004, national consumption doubled in China and tripled in India, while the fastest-growing rates of consumption per person were in Lebanon, the UAE and Mexico.

The phenomenal rise of bottled water sales worldwide spawns a number of social and environmental issues. First, the whole point of bottling water is to export it from an aquifer or water catchment, which contributes to drying local wells and wetlands. Second, bottled water may or may not be safer to drink than water from taps, wells, rivers or lakes, but it is always much more expensive. If national élites can afford to drink safely from bottles, while the poor have to use taps and wells, then some of the motivation for societies to clean up public water supplies is removed. Third, up to three million tonnes of plastic are used to bottle water each year, most of which could be recycled but more than 90 per cent is not. Fourth, making that much plastic uses at least a million tonnes of oil, and the plastic in each one-litre bottle uses five litres of water in its manufacture. Fifth, while tap water flows through pipes in an energy-efficient way, bottled water is transported by sea, rail and road over long distances, which involves burning a huge amount of fossil fuel. Finally, meeting the UN goal of halving the proportion of people who lack a secure water supply by 2015 would need an extra investment of US$15 billion per year, so spending US$100 billion a year instead on bottled water seems rather perverse.

These concerns have prompted calls for consumer boycotts of bottled water, or even for the prohibition of its sale and use. In 2006, the mayor of Salt Lake City in Utah asked city staff to use tap water instead of bottled, and in 2007, the city authorities of Charlottetown in Prince Edward Island, Canada, and San Francisco in California both decided to prohibit the use of bottled water in municipal offices. Meanwhile, a number of church groups in the USA and Canada have called on their members to consider the ethical implications of using bottled water, and the Danish government decided to charge deposits for water bottles from November 2007, which should reduce waste and the impacts of manufacturing and transport. But the key point about bottled water is its effect on the availability of safe water for all. From that point of view, it would make more sense to tax bottled water as a way to finance the world's safe water programme.

COUP DE GRÂCE

Once an aquifer is being thoroughly over-exploited relative to its natural recharge rate, it can be finished off by reducing that rate. This can be caused by anything that interferes with the way that water soaks into the ground in critical places, those points where water finds its way most easily into the body of the aquifer. This happens when the water is blocked, such as by building waterproof houses, roads and car parks in the area. Or when evaporation rates are increased, such as by planting thirsty crops like cotton, or thirsty trees like eucalypts. Or when the swamps and other wetlands that naturally form there are drained for farming. It can also happen when the flow of water off the critical place is speeded up, such as by straightening rivers, preventing floods, or by using impermeable culverts to drain the land. These all prevent water from sinking into the ground, where some of it would normally help to recharge the aquifers beneath. The combination of over-pumping and blocking recharge can cause an aquifer to be completely 'de-watered', as the hydrologists put it.

The changes that kill aquifers can creep up unnoticed, revealing themselves only through subtle alterations in land use. In 2007, I analysed changes in land cover between 2000 and 2006, in Yogyakarta province of Java, Indonesia. I found that the area of forest had fallen by 53 per cent, underbrush by 65 per cent, swamp by 93 per cent and traditional cultivation by 60 per cent, while the area of plantation had increased by 61 per cent and that of settled areas and roads by 45 per cent. This showed a rapid, unplanned change from a landscape of natural ecosystems and traditional farms, towards commercial plantations and urban in-fill. In this process, the last natural vegetation was being removed, costing wild animals and plants their final refuge. Diverse farming systems were being replaced by commercial cultivars, with further loss of biodiversity among traditional varieties of crops. This simplified environment is more vulnerable to pests and diseases, and has a reduced capacity to absorb floods and resist soil erosion and landslides. At the same time, more of the land is sealed by concrete and tarmac and thus unable to absorb rain water, starving aquifers and increasing the likelihood of flash-flooding, while urban pollution sources expand and are more likely to contaminate ground and surface waters.

As an aquifer is drained, the spaces once occupied by water are filled by dust, sand and rock fragments pressed down from above, and the structure of the aquifer collapses. This has two consequences. One is that the ability of the water-bearing layer to hold water is greatly reduced, so the aquifer is irreversibly damaged and incapable of being recharged. The other is that the land surface will subside, exactly like earth over a grave, which sinks as the coffin and cadaver below collapse and decay. A city built on the subsiding grave of an aquifer will sink and crack, and heavy rains will flood it. Thus, parts of at least fifty Chinese cities, including Beijing, Tianjin and Shanghai, have subsided by up to several metres, as the aquifers beneath them are drained and crushed by the weight of buildings. As a result, the authorities in Beijing are having to re-build its inner ring road, and to install systems to give early warning of collapse for structures that are being built for the 2008 Olympic Games. Ground cracks threaten Beijing International Airport, and surface levels and angles have changed enough to endanger the Beijing–Shanghai railway, and to affect the flow of canals needed for water diversion projects.

So too with Bangkok in Thailand, and north Jakarta in Indonesia, where subsidence prompts massive flooding by rain, river and sea water, damage to the foundations of buildings, roads, bridges and buried pipelines, the trapping of sewage in the streets, and the intrusion of salty water into the upper aquifers. And likewise in Mexico City, where ground water pumping started early, and where the centre of the city has fallen by an average of 7.5 metres in the last 100 years, causing extensive damage to foundations and the sewer system. Finally, according to the US Geological Service, major areas of subsidence caused by the collapse of over-pumped aquifers include 36 valleys in California, Arizona, New Mexico, Nevada, Idaho and Colorado, as well as towns such as El Paso in Texas, Baton Rouge and New Orleans in Louisiana, Savannah in Georgia, Williamsburg and West Point in Virginia, and Atlantic City in New Jersey.

TRUTH AND CONSEQUENCES

No one is really *trying* to deplete aquifers, but they might as well be. By 2005, the countries that were known to be pumping water out of aquifers significantly faster than they were being recharged included China, India, Iran, Israel, Jordan, Mexico, Morocco, Pakistan, Saudi Arabia, South Korea, Spain, Syria, Tunisia, the USA and Yemen.

In China, under Hebei province the water table is falling by nearly three metres each year, and even faster in some areas near cities. This is in the heart of the North China Plain, which produces over half the country's wheat and a third of its maize. The shallow aquifer in the region is largely depleted, and well drillers have turned to the deep fossil aquifer beneath it. Wells now have to be drilled to a kilometre deep around

Beijing before reliable fresh water can be found, and wheat farmers in some areas are now pumping from a depth of 300 metres. This makes water increasingly expensive, sometimes too much so to be used for irrigating crops. This has contributed to a shrinking of China's wheat harvest from 123 million tonnes in 1997 to 95 million in 2005, and its rice harvest, from 140 million tonnes to 127 million over the same period. The overall *decline* in China's annual grain harvest since the late 1990s exceeds the total amount of grain produced each year in Canada.

In India, the ground water revolution since the 1980s has seen at least 21 million wells drilled, with a million more added each year, and a doubling of land irrigated by them, from 20 million to 40 million hectares. About 250 km^3 of water is extracted each year, 40 per cent more than is replaced by rain. Hence, inevitably, water tables are falling in most of the country: in Gujarat by between 6 and 30 metres each year, with boreholes even 400 metres deep running dry. In Tamil Nadu, wells are drying up daily across the state, and half the irrigated land has been lost in a decade. In Maharashtra, deep wells have taken ground water for sugarcane plantations and public wells have run dry, so many people now depend on tanker deliveries. In Madhya Pradesh, the once water-rich Malwa plateau is now dry, and nine-tenths of the wells drilled in the last decade no longer function. National food production is not yet faltering, but this surely can't be far off.

Meanwhile, aquifers in Iran are being forced annually to yield five cubic kilometres more than recharge, water tables are falling, and 'water refugees' are starting to be seen. Farmers in Saudi Arabia used up more than half of a ground water reserve of close to 500 km^3 to produce wheat at subsidised prices, in the process also draining the Disi aquifer that the country shares with Jordan. In water-stressed Yemen, the water table is falling by 2–6 metres each year across the country as ground water is taken out of aquifers far faster than recharge. In Israel and Palestine there is competitive pumping of aquifers by Palestinian and Israeli stakeholders, but the rules of access are biased in favour of the latter, not least by the fact that many Palestinian wells are now on the wrong side of Israel's new security wall. In Pakistan, even though most irrigation water comes from the Indus River, which is itself fast drying out, water tables are falling at a rate of several to many metres annually, and ground water has been exhausted in many areas.

In Mexico, 80 per cent of all water is used by agriculture, and ground water contributes 40 per cent of it. Irrigation supports more than half of all farm production and three-quarters of exports, most of which are water-intensive fruit, vegetables and livestock for the US market. As a result, water tables are falling fast in such states as Guanajuato and Coahuila de Zaragoza, and more than a hundred of the country's 653 known aquifers are over-exploited. In Sonora state, the Hermosillo coastal aquifer yielded water from 11 metres down in the 1960s, but this must now be pumped from a depth of 135 metres. Over-extraction has led to sea water intrusion, so agribusinesses are moving inland from the coastal areas, in search of new water sources. Over-extraction of ground water is encouraged by electricity subsidies that cost about US$700 million a year, with the largest farms receiving the greatest benefits. This maintains an artificially high demand for water, and discourages efficient irrigation and the use of less thirsty crops.

Thus a worldwide pattern of aquifer depletion has accompanied the ground water revolution, which is bound to cause water shortages in future where they are not already being felt. After reviewing the dismal global scene, the following point was made by Lester Brown and his colleagues, in the *Encyclopedia of Earth*. They wrote in 2007: 'Since the overpumping of aquifers is occurring in many countries more or less simultaneously, the depletion of aquifers and the resulting harvest cutbacks could come at roughly the same time. And the accelerating depletion of aquifers means this day may come soon, creating potentially unmanageable food scarcity.'

THE WATERS OF THE SAHARA

Unimaginably slow geological movement, where the speed of a fingernail growing is fast, interacts with other events on a shorter time scale – natural climate change, for example, where in mere thousands of years glaciers can advance or vanish, sea levels can rise or fall, and deserts can expand and contract by millions of square kilometres. The Sahara Desert has had many wet and dry periods over the last few million years. The most recent changes are the best known, with the Sahara very dry and vastly expanded 18,000 years ago, and much wetter and smaller 10,000 years later. Then, it was inhabited by people who recorded on cave walls their hunting of elephants, rhinos and giraffes. Even in ancient Greek times, mountain ranges deep in the Sahara were known to be peopled, and North Africa remained wet enough to be one of the granaries of the Roman Empire centuries later.

But the desert must have bloomed in much earlier times too, for hundreds of metres beneath two million square kilometres of what is now the driest part of the Earth's surface lies an immense pool of sandy rock saturated with fresh water. This is the Nubian Sandstone Aquifer System, and it contains around $100,000\,km^3$ of ground water, as much as all the world's lakes, rivers and swamps combined. Some of this water is believed to be close to a million years old, but other dated samples show other ages down to 50,000 years, revealing a succession of recharging floods.

The aquifer was discovered in the 1950s, when Libyan exploration wells unexpectedly spewed water instead of oil. As the scale of the find became clear, this was big news for the dry countries of the area, including Libya, Egypt, Chad and Sudan. But Libya was most active in exploiting it, by creating a huge irrigated farm around the Al Khufrah Oasis. The country then built the Great Man-made River – pipes designed to extract up to 6.5 million cubic metres per day, or over $2\,km^3$ per year, for irrigating crops in northern Libya. Egypt plans to use half a billion cubic metres of this ground water annually.

More recently, it was found that a dry lake bed in the Darfur region of Sudan had soaked away in the last 5,000 years, contributing its waters to the Nubian aquifer or one nearby. This is seen by aid agencies as a water source for the 2.5 million people driven from their homes by fighting and ethnic cleansing in Darfur since 2003. But the existence of the water itself, along with oil, might well have helped prompt this murderous land grab in the first place. According to a June 2007 UN Environment Programme report, deteriorating environmental conditions over many years have helped destabilise Sudan, and have prompted conflict and irreversible population movements in the Darfur region.

This is a microcosm of the water and habitat crisis affecting much of north-east Africa and the Sahel, which features degrading lands and deserts spreading southwards by about 100 km over the past forty years. These changes are linked to overgrazing by a livestock population that has risen from 27 to 135 million animals, and deforestation, with the loss of two-thirds of Sudan's forests in 1972–2001. These problems are widespread in Sudan, and there is mounting evidence of long-term regional climate change, with rainfall in northern Darfur having fallen by a third in the past eighty years, and forecast climate change expected to slash food production by up to 70 per cent.

THE WATER WOMEN OF THARAKA

Tharaka in Kenya used to be a horrid place to live, especially for women. It's a hundred kilometres from anywhere, hilly and rocky, with thin soils, and droughts are a way of life. When it isn't blindingly hot and dusty dry, occasional rains lash the landscape into flash-floods and gullies. Then, for a few days, water can be taken from local streams and puddles. But after that, for women, it was back to walking through the thorn-bush for many kilometres, back and forth to the Tana River, returning with a 30-kilo head-load of water. Education wasn't a priority, and only 5 per cent of boys and 1 per cent of girls finished even primary school. Girls were routinely subjected to ritual genital mutilation, and then married off in their early teens. Men were in charge, and treated women without respect, even addressing them as if they were children. Only men could talk first in any meeting, even idle and drunk ones, and could not be challenged by any woman. This is how Tharaka was, in its poverty, in its way of life.

That was 1996, when ActionAid was just beginning to work in the area, starting with a study of the overall situation, and finding the priorities that people expressed when they were asked in private as well as in public. It became clear that the lack of water was perceived as the greatest need, the most acute and urgent problem. So a study was done to find out where there was water, and how it could be brought to the community. A source of ground water was found, beneath the bed of the dry Katse River. Although this was 74 km away, the route was mostly downhill, so the water would flow. In 1998, ActionAid proposed building a pipeline, and a formula was agreed with the people of Tharaka: they would provide labour, and the charity would provide technical advice and materials.

Following the success of a similar project in Ethiopia, ActionAid decided to ensure that women understood, owned and managed the Tharaka project, because it was women and children who had to fetch water for their families. So the Tharaka Women's Water Users Association, or TWWUA, was established. Women were trained on every aspect of the project, technical, managerial and financial. Digging the channel for the pipeline began in 1999, but was very hard going, and it took five years. But in 2004, the pipe was connected to an underground tank which was filled through filter pipes in the dry river bed, behind a sub-surface dam where water accumulates.

The project had a number of impacts, some direct and some indirect. Water was suddenly available within a couple of kilometres of everyone's houses, supplying about 16,000 people through twelve water kiosks, and with connections also to two primary schools, a secondary school, a polytechnic and a dispensary. Partly because girls no longer had to help their mothers fetch water, and partly because schools had been made free in 2002, school enrolment quickly improved. By 2005, completion rates for girls has reached 33 per cent, almost the same as for boys. As women became better informed, more confident and more assertive, and as girls stayed on at school, both early marriages and female genital mutilation declined in frequency and social acceptability.

Women gained in experience through managing the water project. Both the manager of TWWUA, Mary Kangaria, and its treasurer, Mwikali Kirema, were appointed assistant chiefs, in the sub-locations of Gachigongo and Kamaindi. Three other women, all members of TWWUA, stood for civic wards in the 2007 general elections, and hundreds of women obtained identity cards and the right to vote for the first time. As the ActionAid report puts it:

> Although some men were not happy when they initially heard that women would be managers of the project, women never gave up. Their minds opened up as they discovered their potential. The process is a slow one, but dividends have started to come. Each year has come with something new for women since 2005. From being seen as the least in society, women are taking up positions of leadership and contributing to development.

The transformation since 1996 is deep and irreversible, and has only just begun. It's due to a cascade of effects arising from the ActionAid intervention, the determination of the community to bring water to their homes, the involvement and empowerment of local people to solve a local problem and, especially, the efforts and bravery of the water women of Tharaka.

HIGH PLAINS SNIFTER

The High Plains or Ogallala aquifer lies under 450,000 km² of South Dakota, Nebraska, Wyoming, Colorado, Kansas, Oklahoma, New Mexico and Texas. Much younger than the

Nubian aquifer, it contains water that soaked into the ground towards the end of the last ice age, about 10,000 years ago. It's also, compared to the Nubian yard-glass of water, a mere snifter at only about 3,000 km^3. The High Plains aquifer is only slowly recharged by rainfall, and barely at all by its rivers, since many are below its level and actually help to drain it. Most of the aquifer is covered by *caliche*, a hardened deposit of calcium carbonate mixed with gravel and sand. This is formed when minerals are leached by occasional rains from the upper layer of the soil, and accumulate 1–3 metres beneath the surface, creating a waterproof shield on top of the aquifer. Some of the few places where water can get into the aquifer are in small seasonal lake beds called playas, where *caliche* doesn't form. The average rate of recharge to the entire High Plains aquifer is only a little over a centimetre a year. There are around 20,000 playas in the southern High Plains, but many have been destroyed by farming, or have been built over, so natural recharge is becoming even more limited than before.

The High Plains aquifer is shallow, and its water can be pumped from as little as 30 metres down. It was first tapped for irrigation in 1911, and this expanded dramatically from the 1930s onward. Electricity provided to rural farming communities, and mass-produced electric pumps, ensured that the use of ground water would grow. The High Plains were quickly transformed into a vast and highly productive cattle ranch and a maize, wheat and soya bean plantation. But it couldn't last. The water table was falling at a rate of 30 cm a year in the 1940s, and up to 1.5 metres a year in the late 1950s. Sustained pumping eventually lowered the water table by as much as 30 metres in parts of Texas, where in places the aquifer has been completely de-watered. Over-exploitation started slowing in the mid-1970s, but droughts since the 1990s have renewed the pressure.

Meanwhile, water withdrawals have been reduced in various ways. Government incentive schemes have helped to increase the average efficiency with which water was being used, from about 50 per cent in the mid-1970s, to 75 per cent in the 1990s. Also, less land is being irrigated, since some has been set aside for conservation, while energy costs have risen and farm prices have fallen. The rate of water decline in the aquifer is now more-or-less constant, but the water table does continue to fall. Similar stories could be told of the Central Valley aquifer in California, where water is taken 15 per cent faster than recharge, and the South-West aquifer under Arizona and neighbouring states, where extraction is 100 per cent greater than recharge. The combined over-pumping of the High Plains, Central Valley and South-West aquifers amounts to about 36 km^3 each year.

THE GREAT ARTESIAN BASIN

Australia's mostly a dry continent, most of the time, and it's been getting drier where most of the people live. For fifty years there's been a trend towards increasing rainfall in the north of the country, and declining rainfall in the south-east quarter that includes most of the country's farms, as well as Sydney, Canberra and Melbourne, and in the south-western corner, around Perth. This trend has become harsher since 2001, and in 2006 the south-east recorded less than 60 per cent of its long-term average rainfall. Some areas had their driest year on record, including key catchments of the Murray River, and the Western Australian coast. This consistent trend, which continued throughout the El Niño events of 2002/03 and 2006/07, has raised fears that south-eastern and south-western Australia may be starting to become uninhabitable. Emergency restrictions on water use have been put in place in these regions, including strict limits on the use for irrigation of what little water remains, and plans to build desalination plants are being developed or implemented in Perth, Sydney and Melbourne.

The interior of Australia has far less rainfall than the south-eastern corner, but much of it sits on top of the Great Artesian Basin. An artesian aquifer is one in which the water-bearing layer is compressed beneath an impermeable layer of rock, so that when punctured by a drill, the water is released under pressure, without pumping. Hence the first drilled boreholes in the 1880s discharged high fountains of water into the air. The Great Artesian is the largest and deepest such aquifer complex in the world, underlying 23 per cent of a whole continent, including most of Queensland, the south-eastern part of the Northern Territory, the north-eastern part of South Australia, and northern New South Wales. Its discovery in 1878 led to the expansion of ranching into vast new areas. It is estimated to contain about 8,700 km^3 of water, of which about 0.6 km^3 is used each year. The water's high sodium content makes it unsuitable for growing crops, but with so few surface water sources in this arid region, it's vital for watering livestock and to support mining and tourism.

Wells drilled into the Great Artesian were allowed to run continuously, but the flows of water began to decline after a time. Even before the end of the 1880s there was concern about the waste of water, and a law to ban it was passed by Queensland's Lower House of Parliament in 1891. This was rejected by the Upper House, though, on the grounds that the outflow from wells was insignificant compared with the recharge of the aquifer by rainfall. For it was believed that the Great Artesian is recharged by rain that falls on the Great Dividing Range along its eastern margins, and then flows slowly south and west. This is still the prevailing assumption, but some have questioned whether this theory of recharge and flow makes sense, given the geological structure of the area and the flatness of its terrain. An alternative view, advocated by L.A. Endersbee, is that recharge is by seepage from deep within the Earth, the water having been there since the original formation of the planet. If this were to be confirmed, then many theories of ground water recharge would need to be re-examined.

EARTH TAINT

It isn't just the quantity of water that's important. Pollution can reduce the availability of fresh water as surely as drought, and this pollution can have many sources, including the Earth itself. Even though hundreds of millions now depend on ground water for drinking as well as irrigation, not all of it is actually safe to drink. This is because, during its long travels and residence underground, the powerful solvent that is water picks up and dissolves many chemicals. These vary by location and depth, but can include the radioactive gas radon, sulphates, chlorides, fluorides, nitrates and the elements selenium and arsenic. Deep wells, which are increasingly being used as shallow aquifers become exhausted, are most likely to yield radon and fluoride derived from granite rocks.

In the 1980s, the UN Children's Fund, or Unicef, along with other donors, provided large numbers of deep wells to villages in central and southern India. Unanticipated and untested for, the water they yielded contained up to fifty parts per million of fluorides in some cases and typically around twelve, which is still ten times the maximum concentration recommended by the World Health Organization. At these levels fluorides are poisonous, and with long-term exposure they cause crippling bone deformities and bone weaknesses that lead to hip and wrist fractures, as well as anaemia, stiff joints, kidney failure, muscle degeneration and cancer. Tens of millions of people now have such symptoms in seventeen Indian states, of which Uttar Pradesh, Rajasthan, Gujarat, Andhra Pradesh and Tamil Nadu are worst affected. In Rajasthan, more than half the population has symptoms. Since this disaster, dangerous levels of fluorides have been found in wells in Africa all the way down the Rift Valley from Eritrea to Malawi, and across Asia from Turkey through Iraq, Iran, Afghanistan, India and northern Thailand to China, with China alone having a million cases of chronic fluoride poisoning.

The same rush to provide well water for the poor in the 1980s led to another accidental mass poisoning, this time involving arsenic from shallow wells in Bangladesh, and in the Indian state of West Bengal. The first 900,000 wells were again provided by Unicef, but other donors also participated. The wells were not tested for arsenic until years later, when doctors noticed large numbers of people with unusual symptoms. About 60 million are now drinking ground water with arsenic concentrations in the range of 50–2,000 parts per billion, in places even higher, well above the WHO limit of 10 parts per billion. At these levels of exposure there's chronic damage to the human body. This includes changes in skin colour and texture, and an increased chance of cancer in the skin, lung, kidney, bladder and lymphatic system. An additional 300,000 deaths from cancer are expected from chronic arsenic poisoning over the coming decades. People in many other countries are also affected by arsenic in well water, including in Cambodia, Nepal, Tibet and Vietnam. So, when you're asked to help raise funds for drilling wells in poor countries, it's important to ask the charity what they know about the safety of the water they're drilling into, and what they're doing to protect the level of the water table that they're drawing from. If the answer is 'don't know', don't give.

THIRSTY FARMS

The sound of Bali is the gentle tinkling of water running in bamboo pipes among rice fields, dribbling from terrace to terrace. A thousand years of detailed understanding of how to maintain harmony among soil, crops, spirits and water is at work here. To find something similar elsewhere we must turn away from the vast and mechanised plantations of the world, where soils are ruthlessly depleted, crops are fertilised and sprayed with petrochemical products, and water is swilled, wastefully and destructively, across the landscape. We should turn perhaps to a small organic farm in England or India, where the chemical complexity of soil humus is cherished along with all its microscopic biodiversity. Where air and water are seen as all that a good soil needs to produce abundant and excellent food. Where rain-water capture and small-scale irrigation using drip-feed techniques are used. Where mulching and inter-cropping replace artificial fertilisers, and natural predator-prey relations are relied upon to replace pesticides. And where traditional farming practices and systems are valued, preserved and restored.

But these places are exceptions rather than rules. The recent past has mainly been about farming ecosystems within an inch of their lives as the human population has rocketed, and demand for food and all other elements of survival, commerce and prosperity has risen with it. Fields have been flooded wholesale with irrigation water, promoting waterlogging and salinisation. Thirsty cultivars that require irrigation have been developed and distributed. Production increments have been forced by using large amounts of artificial fertilisers and pesticides at the cost of soil biodiversity, chemistry and structure. Farming systems have been simplified away from those with small fields, hedgerows and multiple crop species, in favour of large-scale monocultures. And imported water has made much of this possible.

WATER BALANCE

Using imported water for farming, whether it's channelled from far away or pumped out of the ground, breaks the links between ecology and human livelihoods. In the past, the crops that people grew in an area were strongly influenced by its climate, and the issue of growing thirsty crops like cotton, alfalfa or sugarcane in a dry zone didn't arise because we just couldn't. So, among all the issues that arise from irrigation and the ground water revolution, from aquifer depletion to salinisation and mass poisoning, is that we've largely suspended the rules of *carrying capacity* as they apply to us. This is the idea that for any species in an ecosystem, numbers are limited by the resources produced by that ecosystem under prevailing conditions. By introducing water to an arid ecosystem, we've removed one of the key constraints within that ecosystem, a basic factor that determines its nature. In short, we've broken it.

Now we grow the wrong things in the wrong places, and plaster over the ecological question marks with irrigation water, agrochemicals, government subsidies and genetic engineering. We produce milk in dry areas with irrigation water, transport it around Europe, and wreck the UK's dairy farms, despite the fact that the UK's naturally moist grasslands are most suitable for grazing animals. We are proud that we can make holes through mountains, for example to divert 2.1 km^3 a year or 99 per cent of the Snowy River's flow from one side of Australia's Great Dividing Range to irrigate crops on the other. We allow a coal company to take 4.5 million tonnes of high-quality water each year from the aquifer that sustains Hopi communities in Arizona, in order to pump coal slurry to power stations in Nevada. And we grant electricity subsidies to large agribusinesses in Mexico and sugar barons in India, so that they can drain aquifers and impose ground water famine on their weaker neighbours. The impacts of such choices on livelihoods are just as profound and irreversible as they were for the fishing people of the Aral Sea when Stalin's cotton fields got top priority in the USSR. Meanwhile, globally, we continue to raise the stakes in all directions, packing in more of everything, demanding more of everything, taking no precautions and no care. But the rules of carrying capacity can't be suspended for ever.

GROUND WATER CHOICES

We've seen how we can wreck aquifers by accident, simply by forcing them to produce more than is in their nature to supply. In similar management choices, we can subsidise access to drills for wells and electricity for pumping water, and we can artificially support the prices of thirsty crops or use them in inefficient ways, such as for cattle feed. We can develop boreholes as hasty acts of charity, without contemplating sustainability or testing for naturally occurring poisons such as fluoride and arsenic. We can invest in or allow large-scale aquifer pumping to feed cities, industries and irrigated plantations, neglecting the land subsidence and irreversible growth in demand that will result. We can privatise wells and water supplies, so that a lack of equity is added to the brew. And we can prevent the recharge of ground waters by covering land with impermeable roads and towns.

But should we wish to take a longer view, there's always the more ecological, 'Taoist' alternative. We could encourage people to construct or restore check-dams, ponds, tanks and diversions of seasonal rain water into wells, so as to promote ground water recharge. We could stop buying bottled water and insist that our water suppliers provide high-quality water to our taps. We could limit ground water extraction to natural recharge rates. We could protect wetlands and design and build porous towns and cities, with enough green spaces for ground water to recharge. We could insist on labelling to tell us the virtual water content of everything we buy. And we could grow crops and animal products that match the sustainable local water supply, rather than base agricultural policies solely on profit or the protection of special interests.

9. THE WORLD TO THE RESCUE?

I had lunch with Betty, an eighty-year-old lady who's interested in the environment. We talked about water and rainforests, biodiversity and the future. At one point, she looked thoughtful, saying, 'My lot have done this to the world and it's too late for us to undo it. I do worry about my children, though, and their children. I hope you and they'll make it better.' I'd put low on my blame list people born in the 1920s, who spent their lives in the Women's Institute, delivering Meals on Wheels, and volunteering for good causes. We might instead look closer to home, at the generations who drove the explosion of the world's economy since about 1950. As a result of this explosion, we now appropriate at least 40 per cent of the entire photosynthetic production of the Earth, challenging all other species to survive on the remainder. The list of other impacts is a long one. But it's the same people, and their children, who'll solve the problem so, while I'm all for learning from the past, saying sorry and doing something about it, blaming people isn't going to help. Instead, the first step is to have a look at the scale of the problem.

PROBLEMS WITH WATER

Water rations

If you have to search long and hard to find a house with reliable water, as I once had to do in Africa, it implies that there are plenty of other houses without it, where less lucky people have to live. And when you see that, you notice that the whole world is like this: some have baths and well-watered farms, but many don't. And it turns out that 'many' means hundreds of millions, even *billions* of people. And they aren't getting any fewer. This is worth thinking about, for in these last few decades we've drifted into a global water crisis, an emergency focused on fresh water without sewage or poison in it. That's getting rarer by the day in most of the world, and 40 per cent of all people are now facing serious shortages, a figure likely to rise dramatically and soon. Over a billion people lack access to adequate water supply and more than 2.5 billion have inadequate sanitation, causing millions of illnesses and deaths every year, mostly among children.

A share of the relatively tiny amount of liquid fresh water, in lakes, rivers, swamps and clouds, is needed by every person, and every one of the millions of species, on Earth. We now use over $3,000 \text{ km}^3$ of fresh water each year, which is twice as much as the entire volume of Lake Ontario, or 8,700 times the volume of Windermere. With this astronomic and increasing demand, supplies are falling so that, if you are twenty now, when you reach forty your average ration will have dwindled by a third. That is, if you can get your lips to an average ration, and most people will not, since competition for water is growing fast where most people live. Irrigation takes some 70 per cent of all fresh water used by people, and this demand is rising, but industry's needs are also growing, and are expected to reach a quarter of all fresh water by 2025. That would leave nothing at all for domestic use in cities, where half the world's population already live, and where two-thirds will live by 2050. When big businesses and rich people need water from a limited supply, it will tend to be the poor and vulnerable who lose out, with dire consequences for their well-being, health and livelihoods. The net result is that the water situation is becoming critical in many places, and would be alarming even if climate change weren't making the whole thing worse than it already is.

Water scarcity

The abundance of water isn't an absolute thing, only a relative one. It's scarce only in relation to what ecosystems must have before they transform themselves, whether these are natural ecosystems, like wetlands, or artificial ones, like farms. Likewise, it's only scarce relative to the demands of cola bottlers, car factories, power stations, plantations and cities, not in and of itself. Natural deserts are just as perfect ecosystems as rainforests, and all peoples who know how to use the water they have, and no more, are equally perfect dwellers on the planet. Problems arise with expectations, either when cultures and economies adapted to one kind of water abundance relocate to another place, and build a city like Perth in south-western Australia, or when ideas spread and encourage new visions of future lifestyles and activities that cannot be sustained in their new homes. Or when new technologies, such as deep drills and power pumps, temporarily make possible new settlements and farms. Or when new markets demand products that take a lot of water to make, such as biofuels, and force the opportunistic development of new plantations in irrigated areas. Or when old ideas of water conservation become seen as outdated, and are forgotten. Or all of the above. At which point we turn to our engineers to solve the problem, rather than our ecologists.

Water distribution

Fresh water is not distributed evenly around the world, and for the most part this cannot be changed. Most of the 40,000 km^3 of flowing river water in the world at any one time is located in Siberia, Canada, and the Congo and Amazon basins, far away from centres of human population. Huge engineering schemes can rearrange things a bit, but at a cost. We've seen how in the former Soviet Union, for example, 40–50 km^3 of water were diverted each year from rivers feeding the Aral Sea to new cotton farms in distant deserts, but the price was the Aral Sea itself. New schemes in China and India aim to move similar amounts of water each year from the wetter parts of their countries to the drier ones (the north in China, the south in India). These will have their own ecological impacts, with losers as well as winners, both human and non-human.

The most effective way we've found to move water around the world is actually in the form of 'virtual water', in which water 'moves' from wetter areas to dry ones in the form of goods that are manufactured and foods that are grown using water, at a rate of 1,000 km^3 each year. Dry countries can thus survive by using virtual water, and by reserving real water only for drinking and washing, but only if they have something to sell in return, such as oil. They can supplement their real water supplies too, from aquifers (if they have them) or desalination (if they are by the sea), but only if they can afford the energy to pump and process the water. Countries that are poor and dry will suffer worst and soonest, being joined later by those that are dry and become poor, because they've nothing left to sell, or that become dry because they've mismanaged their water-bearing ecosystems or are subject to a drying climate. Thus the water crisis will chew its way up the chain of nations, from Yemen to Australia.

Water quality

Nibbling away, sometimes biting, at the margins of fresh water supply is pollution. Salts, of course, especially sodium chloride but including scores of others, are what make 97 per cent or so of the world's liquid water useless to us for drinking and farming, and barely more useful for washing our bodies and clothes. This salty sea water can also get into our precious fresh water supplies, by intruding into ground water near coasts, if we take out too much fresh, or by being swept inland by storms and tsunamis, or even blown there as spray from crashing waves. The same and other salts exist in the ground too, having accumulated there by geological processes. These we can access, should we be careless, by drilling ground water out of rocks naturally contaminated by arsenic or fluoride. Or we can concentrate them by means of irrigation methods that draw them to the surface through waterlogging and evaporation, or by diverting water through soils before letting it drain back into the river bearing a new load of salts. Thus, people can squander fresh water using only the natural chemicals of the Earth.

But we can do worse than that, much worse. We can flood our rivers accidentally with cyanide, as in the Tisza in Hungary in 2000, or with pesticides, as in the Slea in Lincolnshire in 2003, or with benzene, as in the Songhua in China in 2005, or with raw sewage, as in the Hudson in New York in 2007. Or we can let pesticides and fertilisers leach from our farmlands, and mercury, arsenic and cadmium leach from our mine tailings. Or we can pile our solid wastes in the countryside, for instance in China, where in 2005 more than 1,300 km² of farmland were reported to have been ruined by solid wastes, and in insecure landfills, so that black and stinking leachate oozes into aquatic ecosystems. Or we can bury our dead impregnated with formaldehyde, so that ground water coagulates the proteins of living things that pass through our cemeteries. And in our increasingly populous cities, we can allow day-to-day life among the urban poor to become ever more dominated by the scarcity of safe drinking water, and by escalating concentrations of sewage, industrial effluent and garbage leachate in what fresh water there is.

Ecosystem change

Farms, gardens, plantations, cities, forests and seas are all ecosystems, so agriculture, forestry, wildlife and fisheries management (and much else) are all basically about managing ecosystems. From a human point of view, important features of ecosystems include their ability to respond flexibly to new conditions, such as those imposed by climate change and new demands by people. Equally important is their ability to buffer and aid recovery from stresses and shocks, such as droughts, floods, pollution, fires and wars, and their ability to supply ecological goods and services. Most vital among these are water supply, stability and purity, while in farming such services also include pollination and pest control, and in fishing they embrace feeding and sheltering fish while they breed and grow. In particular, the regular supply of fresh water depends on healthy ecosystems, especially catchment forests and wetlands, so it's not surprising that crises of water scarcity, distribution and quality are hitting us now, after decades spent abusing them worldwide.

Natural forests, the ones made of native species, have been catastrophically reduced in many countries, and widely lost, fragmented and degraded elsewhere. Meanwhile, as we saw in previous chapters, local decision-makers have often fallen into the trap of seeing wetlands as cheap, reclaimable or disposable wastelands, and these ecosystems have therefore been devastated by drainage, dams, pollution, construction, farming and fire. Lakes have often been seen only as large bodies of water that can be used profitably for irrigation, with catchments suitable for farming, logging and settlement without attention to the one-way flow of agrochemicals, sewage and eroded silt into the lake, or its sump-like vulnerability to accumulating toxins. Rivers have been so treated, with the ecological consequences of human use multiplying downstream, that the final users have to contend with eroding estuaries, destroyed fisheries and devastating floods, or, at other times, a trickle of salty, muddy and toxic water. Finally, the local balance of ground water has often been affected by preventing floods, by waterproofing or draining the surface artificially, by changing the vegetation, or by irrigating.

WATER AND THE WORLD

Protecting the oceans

There is a global treaty on the oceans, of a sort. Called the United Nations Convention on the Law of the Sea (1982), it became binding internationally in 1994. Commonly called UNCLOS, it tidied up the results of a series of previous UN conferences and treaties. In the 1950s, these had addressed governments' rights and duties concerning their territorial seas and continental shelves, and in international waters. Follow-up meetings in the 1960s fell foul of the Cold War, with participants congealed around the incompatible positions of the USA, the USSR, and their clients and satellites. Meanwhile, other efforts to protect the oceans included the Convention on the Prevention of Marine Pollution by Dumping of Wastes and Other Matter (1972), and the International Convention for the Prevention of Pollution From Ships (1973). Encouraged by these, there was further progress on UNCLOS in a series of meetings from 1973 to 1982, during which agreements were hammered out on the various kinds of sea, legal rather than ecological, and the rights that governments have in them.

These legal zones are defined relative to a baseline close to shore. They include internal waters (inside the baseline), territorial waters (out to 12 nautical miles), contiguous zones (out another 12 miles), exclusive economic zones (out to 200 miles), archipelagic waters (for countries made up of islands, like Indonesia and the Philippines), and the continental shelf. In all these zones, the owning government has complete discretion with regard to the exploitation of marine and sea bed resources, but other stakeholders may have certain rights, such as 'innocent passage' (which excludes fishing, polluting, testing weapons and spying).

Beyond the continental shelf are international waters, or High Seas, which can be claimed by no one, but here Part XI of UNCLOS established a new International Seabed Authority to authorise sea bed exploration and mining, and to collect and distribute mining royalties. This new mechanism was activated in 1994 but is a major sticking point for the USA, which signed but has never ratified the convention, although it accepts most of its other provisions as binding international law. These provisions include general obligations for safeguarding the marine environment, and protecting freedom of scientific research in international waters.

The oceans provide a backdrop to the fresh water issue, since they do so much to control the weather and climate, and ultimately provide fresh water through evaporation and rainfall. They are the most influential of all the adaptive systems of the biosphere, and the ultimate source of all life. Their fate will help determine our future, through the relationship between our land-based activities, the composition of air and sea, and the overall response of the biosphere to solar radiation.

Though vast, the oceans still have localities and local ecological challenges. Dead zones are caused by particular kinds of pollution, mostly created on shore, which may be concentrated by currents, or the lack of them, to a point where algal blooms kill all the sealife in an area. Similarly, the populations of particular fish species don't extend evenly throughout the oceans but are instead divided into geographical units (like the orange roughy populations of New Zealand and the UK), or are managed as if they were (such as the Iceland and North Sea cod populations), or else are associated with geographically limited features like upwellings and sea mounts. Likewise, national or community fish sanctuaries can be established, and these will lie within designated areas and contain a particular set of local ecosystems. In all these cases, it is the decisions of local environmental managers that influence events, for good or ill, in local parts of the ocean ecosystem.

There is an interface between local decision-makers and the international community in the form of the Global Programme of Action for the Protection of the Marine Environment from Land-Based Activities. Usually known as the GPA, and with a secretariat provided by the UN Environment Programme since 1995, this is the only inter-governmental programme that addresses links between fresh water and marine environments. It's based on the idea that, with a billion people living in coastal cities, and 80 per cent of ocean pollution coming from the land, major threats to the health of the oceans must be addressed on land. This approach has been endorsed by 108 governments and by the EU, which all participate in the GPA. It mainly involves sharing information about how governments and other stakeholders can meet their obligations under UNCLOS and other international laws, and their various policy commitments to protect and develop sustainably the resources of the planet's marine and coastal environment.

Towards a global water treaty

There is so far no global agreement or treaty to establish and codify our various roles, rights and responsibilities with respect to fresh water. This is being blocked by a lack of consensus. The United Nations seems clear that 'Access to safe water is a fundamental human need, and therefore a basic human right', according to former UN Secretary-General Kofi Annan, and there are many who agree. This approach certainly validates global investment in water and sanitation systems to meet the needs of deprived peoples. Such investment could be in the form of tax-funded public giving, which would be straightforward but expensive. Meeting the UN goal of halving the proportion of people who lack a secure water supply by 2015 would need an extra investment of US$15 billion per year. On the other hand, the world now spends US$100 billion a year on bottled water, so perhaps the deterrent cost of public global investment in water and sanitation is rather less financial than political in nature.

In any case, such investment could also be in the private supply of water and sanitation, which many governments prefer as it seems cheaper and more effective, and gets rid of their own responsibility. But if big corporations based in richer countries can do the job, why should taxpayers in the same richer countries compete with them by giving water systems away? The challenge, though, is that for profits to be made, people who use water must pay for it, and without fierce and effective supervision, corporations will tend to charge too much and cut corners. The story of water privatisation worldwide is littered with cases in which the few major corporations involved, and their many subsidiaries, have negotiated monopoly contracts, only to rack up prices and renege on sewage treatment investments. Cases such as this are known from Argentina, Australia, Bolivia, Ghana, Mexico, the Philippines, South Africa and elsewhere. Details can easily be found in some of the books in the reading list at the back of this book, in particular those with titles like *Water Wars* and *Whose Water is It?* Here you can find sentences like Maude Barlow's: 'A handful of transnational corporations are aggressively taking over the management of public water services in countries around the world.'

It must be said, though, that it is *possible* to get private corporations constructively involved, *provided* that water users and local and national governments all keep an eye on them, to prevent profiteering and enforce the terms of agreements, *and* provided that the system is designed with ecology in mind. There's no particular reason why a private corporation, rather than a public utility, should not sell clean water and sewage services to those who can afford them, *provided* that those who can't pay also have reasonable access (which is where a right to water and a targeting of public funds would come in), *and* provided that the private corporation is also held responsible for both the downstream environments that absorb wastes and the upstream ecosystems that provide the water. It is rare but feasible for a city administration to arrange for water charges to be shared with local people who live in and around water catchments upon which the city depends, in return for safeguarding the forests there. In this context, forest and catchment managers are just as much employees deserving of a fair wage, as anyone else who produces something that society wants or needs.

The challenge, however, lies in the process of negotiation, education, consensus-building and legislation needed to make it happen. This will always be a local process, dealing with local ecosystems and local stakeholders, and finding local, fair and sustainable solutions. But a global agreement could validate the idea, and encourage knowledge-sharing about what's possible, what water is worth, where it comes from, what kind of contracts might be needed, and other useful, essential, but frequently overlooked practical details. At any rate, I'd rather see a fresh water treaty like that, spelling out how human needs are to be met in practice, and linked to the conservation of real ecosystems, than one filled only with good intentions and meaningless targets.

Gold standards for fresh water

Other stumbling blocks to a global water treaty include the sheer diversity of human circumstances – ecological, economic and political. However, Europe shows what can be achieved in a relatively homogenous place, one with a common social system based on decades of consensus building after centuries of civil war. In this sense Europe is unique, but it also gives a hopeful signal that people can get their act together eventually. The first accomplishment was actually brokered in collaboration with the UN, and was the European regional Convention on the Protection and Use of Transboundary Watercourses and International Lakes (1992). This obliges members to prevent, control and reduce water pollution. The EU built on this treaty through the Water Framework Directive (WFD) (2000), which requires integrated river basin management, and aims to ensure clean rivers and lakes, ground water and coastal beaches throughout its member states.

The WFD is a unique 'gold standard' in the management of water resources. It sets standards for river basin planning, and for the ecological quality and chemical purity of surface and ground waters. For river basins, the aims are general protection of aquatic ecology, and specific protection of unique and valuable habitats, drinking water resources, and bathing water, and all these objectives must be integrated for each river basin. The central requirement of the WFD is that the environment must be protected to a high level, in its entirety. For ecological quality, water bodies are supposed to show no more than a slight departure from the biological community which would be expected with minimal human impact – the equivalent, I suppose, of a Canadian lake exposed to summer campers and duck-hunters.

For chemical purity, the WFD requires that surface waters must comply at least with all the quality standards established for chemical substances at the European level, with higher standards for particular zones, while ground waters are, as a general principle, not allowed to be polluted at all. The approach is precautionary, although some standards have already been set for ground water at the European level, for nitrates, pesticides and other biocides, and these must always be adhered to. Using a mixture of absolute prohibitions and standards, and monitoring, reporting and restoration requirements, the WFD aims to ensure the protection of ground water from all contamination. For good measure, the WFD also limits the amount of ground water that can be taken to that portion of the overall recharge that is not needed to support connected ecosystems such as lakes, rivers and wetlands.

The EU has also issued the Urban Waste Water Treatment Directive (1991) and the Nitrates Directive (1991). These together aim to tackle the problem of eutrophication, the accumulation of nitrate and phosphorus compounds from sewage and fertiliser pollution, which causes excessive algal growth and thereby suffocates aquatic life. They also target health issues such as microbial pollution in bathing water, and nitrates in drinking water. The EU's later Integrated Pollution Prevention and Control Directive (1996), deals with chemical pollution. The WFD lays down how the application of these other directives is to be co-ordinated with the implementation of the WFD itself. All these directives are required to be written into the national laws of the EU member states. The EU approach is impressive, and if extended effectively to the world as a whole would go a long way to address the issues and problems outlined in this book.

Conserving ecosystems

Meanwhile, global efforts to manage fresh water have been less direct, and more to do with the fact that water and ecosystems are intimately linked, so their management cannot be separated. Thus the rationale to set aside protected areas has, for decades, included the argument that their role in water supply is vital, even if most of the conservation fundraising and political pressure has historically focused more on the need to conserve wildlife. Hence, the international response to water issues is, even if only accidentally, in large part the same as the global effort to conserve the ecosystems that provide the fresh water upon which we all depend. Looking at the main ecosystems that sustain fresh water supplies on land, there is at least a consensus among governments that they should set up and manage protected areas.

This is central to the aims of the Convention on Biological Diversity (1992), and the establishment of protected areas is a key indicator of our progress in halting mass extinction. Global targets have been set to reduce the rate of biodiversity loss by 2010, either 'significantly' (as agreed at the 2002 Johannesburg World Summit on Sustainable Development), or 'entirely' (the aim of the European Union). These aims cannot be achieved, but it has at last been realised that conservation of ecosystems is vital if anti-poverty goals are to be met, since they sustain livelihoods in ways that cannot be substituted by anything else. Thus, one of the Millennium Development Goals, agreed by a UN conference in 2000, was to 'ensure environmental sustainability' by 2015, by reversing the loss of environmental resources and halving the proportion of people without sustainable access to safe drinking water. The wording is a bit muddled in the original, but the concept that water and environment are deeply linked is there.

The gradual growth in this kind of thinking has been matched by an expansion of the world's protected area system, from 1 million square kilometres in 1948, to 2 million in 1961, 5 million in 1972, 8 million in 1979, 12 million in 1992, and 18 million in 2002. This is the most obvious accomplishment of the conservation movement, which built on the approach pioneered by the earliest national parks, such as the Yellowstone (1872), the first created anywhere, and the Virunga (1925) and Kruger (1926), which were the first and second national parks in Africa. These and other protected areas are all constituted under national laws, but a selection have also been inscribed as natural World Heritage Sites, under the Convention Concerning the Protection of the World Cultural and Natural Heritage (1972), and others are listed as wetlands of international importance under the Ramsar convention (1971). These additional listings are meant to signal the special interest of humanity in their safety, and are supposed to make it easier for them to receive protection and funding.

But the problem with protecting some areas, or indeed listing some as 'extra special', is that it gives the impression that everywhere else is more expendable. This is especially untrue for water, since everywhere is a catchment for somewhere. Hence there have been efforts to use these parks as core areas for landscape-wide environmental management systems that also include buffer zones, corridors between parks, 'peace parks' across national frontiers, community reserves, special wildlife management zones, permanent forest reserves, etc. In several places, countries have collaborated to try to manage whole river basins, including those of the Danube, the Mekong and the Congo, with mixed but generally helpful results. It's important to note that the world's system of protected areas was not established with climate change in mind, and needs to be reviewed – possibly even reorganised – in light of changes to ecosystem and wildlife distributions that are bound to accompany new rainfall and temperature patterns, and new sea levels. The existence of huge areas of protection, and linking corridors among them, will certainly help to buffer the biosphere against the impacts of climate change to some extent. But the questions remain: how much? And will it be enough?

DISASTERS, SLOW AND FAST

Desertification

Natural deserts are ecosystems in which organisms have adapted to an annual rainfall of less than 250 mm, often much less. The struggle to exist in deserts reinforces the idea of symbiosis between life and water. They are diverse and fascinating places, but often devoid of people because of the lack of water and sparse vegetation. Desertification, by contrast, is an insidious and unnatural process that destroys plant communities and soils, thus degrading the landscape to a point where it looks superficially like a natural desert, and is at least as useless to people. It is not always linked to a dry climate, and in the US state of Maine, for example, severe soil erosion, caused by farming and over-grazing in the late nineteenth century, exposed a small area of sand left by a retreating glacier, which is now a tourist attraction. Likewise, in Indonesian Borneo, a vast rice project was located in an area of peat swamp forest in the 1990s, and has left behind some 4,600 km^2 of barren white sand. This had once lain under the waterlogged peat on which the forest had grown, which used to be a rich habitat for orang-utans.

But desertification is more strongly associated with dry areas, and it's widespread around the fringes of natural deserts, such as the Sahel in Africa, where the Sahara is spreading south at about 25 km a decade. It's severe in Afghanistan, Kazakhstan and elsewhere in Central Asia, as well as in western China, the Indian states of Rajasthan and Chhattisgarh, and in Mongolia. Some 10 per cent of the island of Madagascar has been desertified, Nigeria is losing about 3,500 km^2 annually, and deserts are expanding in Brazil and Mexico. The problem is often a savage combination of over-grazing and poor farming practices to expose the soil, drought to weaken it, and wind to blow it away. Much the same combination created the 'Dust Bowl', a catastrophic series of dust storms in US and Canadian prairie lands in the 1930s, in which much of the region's soil was lost to the Atlantic Ocean. The UN Convention to Combat Desertification was signed in Paris in 1994, and aims to combat desertification and mitigate the effects of drought through national programmes supported by global partnerships.

Disaster preparedness

Desertification is a creeping disaster, but there are others that are more instantaneous. Over 300 million people are affected each year by earthquakes, storms, floods and volcanoes, for example. The poorest communities are usually hurt most, because they tend to live in greater densities, in badly built housing, on land at risk. Almost all disaster-related deaths occur in developing countries, and disasters especially damage the economies of the poorest nations. Emergency aid can take days to arrive after a calamity, so it's vital for people to be prepared. In practice, the most effective life-saving efforts are usually made by the affected people themselves, during and immediately after disasters. Seeing this as an important opportunity to improve things, both the United Nations and the European Union have programmes to help people understand, prepare for and respond to disasters.

The UN's version is called Awareness and Preparedness for Emergencies at Local Level, or APELL. It was originally developed for industrial disasters, and later adapted to natural ones. Its purpose is to build the capacity of local emergency services to cope before, during and after disasters, and to raise community awareness about the risks they may face and what they can do about them. Essentially, it helps local people develop the knowledge and arrangements for making decisions to deal with hazards. It's based on a ten-step process in which participants understand and evaluate hazards, think through how they might respond to them, and write a plan that is then used to raise public awareness.

The EU's version also has a convoluted name, one so complicated that it's only ever called DIPECHO. ECHO is the European Commission's humanitarian aid department, and DIP stands for 'disaster preparedness'. It targets vulnerable communities living in the main disaster-prone regions of the developing world. As it's very difficult to prevent or influence natural hazards, the programme concentrates on reducing vulnerability in advance. Its main goal is to ensure that disaster reduction measures are fused with wider national policies, for example on education, building codes and health.

Disaster risk reduction

The World Conference on Disaster Reduction took place at Kobé, in the Hyogo prefecture of Japan, in January 2005, right after the Indian Ocean tsunami. It agreed five priorities for action: to make disaster risk reduction a priority at national and local levels; to identify, assess and monitor disaster risks and enhance early warning systems; to build a culture of safety and resilience at all levels; to reduce underlying risk factors; and to strengthen disaster preparedness. The Hyogo Framework for Action for 2005 to 2015 sets out a framework for national initiatives, and many countries have since produced national reports on how to implement it. Another outcome was the creation of a strengthened UN International Strategy for Disaster Reduction, with a secretariat in Geneva to organise a global forum, and to support national activities in line with the Hyogo plan.

Mangroves for the Future

The Indian Ocean tsunami riveted attention on the role of coastal ecosystems in disaster risk reduction. This joined with the growing belief that climate change is increasing the likelihood of disasters, especially in coastal zones where increasingly fierce storms threaten ever-greater numbers of people. The Mangroves for the Future initiative was therefore set up in 2006 to mobilise funds and expertise to encourage the protection and restoration of coastal ecosystems around the Indian Ocean. It's co-chaired by the World Conservation Union and the UN Development Programme, and links several Indian Ocean countries and various international organisations and charities. Its key focus is to promote the idea that coastal ecosystems are economically important 'development infrastructure', so more should be invested in protecting and restoring them. This emphasis tended to divert attention from the local ecosystem restoration activities that have proven so effective in Indonesia, for example, as described in Chapter 5. This key link may be re-established, though, once the initiative's preparation phase is completed in late 2007.

BENEATH THE GLOBAL RADAR

All this global action is reasonable, but doesn't really help to explain water crises in a way that will lead to more understanding of their causes and the potential solutions to them. That's because most water issues are to do with decisions that affect *local* ecosystems and the use of water in them. The global crisis may exist, but it's rooted in tens of thousands of local crises, caused by millions of local choices within local power structures, by people with interests to promote in competition with others. Inevitably their decisions have tended to be oriented to the short term and immediate economic gain, rather than to long-term sustainability. International attention can really only focus on the apex of this structure, either the big picture of emerging poverty and disease, which is the chief interest of the global aid community, or the resolution of disputes between countries over shared water resources, where dialogue and agreement might in principle (but rarely in practice) be facilitated. Local water management issues fall within the remit of national governments and their agencies, but beneath the radar of international bodies, and local managers often lack a coherent view of the broader situation. So now we'll look at the potential for local actors – you and me and everyone out there in the real world – to make a difference to the global water crisis.

10. PEOPLE TO THE RESCUE!

THE BIOSPHERE, HAVING BEEN SAVED . . .

The year is 2085. On the High Seas, in international waters, ships move under 24-hour satellite surveillance, their location continuously tracked and every movement in their fishing gear, cargo and ballast holds logged and measured. All merchant vessels are double hulled at least, and possess a host of other safety technologies developed during the '1.5-degree' storms that began in the first quarter of the century. The global fishing fleet is greatly reduced from early twenty-first-century levels, its vessels tightly licensed and closely super-vised under agreements among every nation and federated region on Earth. Licence fees and freight taxes help pay for scientific studies and monitoring of all fish stocks and the health of marine ecosystems. The rules are amended from time to time in response to new data, with fine-tuning through informed debate among the parties to global agreements. Policing is thorough and effective, and punishment for cheating is harsh.

Within the exclusive economic zones (EEZ) of each country or federation, national marine ecosystem management policy is enforced in line with agreed global standards. National parks ensure that selected areas of great beauty and special value are preserved, for recreation, environmental security and science, as well as to provide feeding and breeding grounds for fish stocks. The latter are harvested elsewhere in the EEZ under strict national supervision or, closer inshore, by coastal communities which have been encouraged by governments to establish their own management zones. Exclusive fishing areas and marine sanctuaries have been laid down by local councils along tens of thousands of kilometres of coast, safeguarding local resources and ensuring that any profits to be made are local ones. Enforcement in these areas is by local policing supported by national and federal government. Many communities are making full use of their property rights by joining biodiversity prospecting agreements with commercial firms and universities. These are searching for new products and processes among the immense array of evolved chemicals in living systems. Up-front payments and royalties from these agreements and discoveries have already transferred billions of dollars to networks of co-operating towns and villages.

As a result of these new arrangements, and adaptation to climate change, fish stocks are showing signs of recovery over large areas of the sea, and permitted harvests are beginning to edge upwards for some species in several countries. This is starting to relieve prices in city shops, where wild-caught fish from the deep sea, like meat, has long been an expensive luxury. Most people's diets have not suffered from this, since in many areas near-shore fish have always been available from the management zones controlled by coastal communities, and there has generally been a good supply of farmed fish and vegetable food, albeit supplemented at times of famine by emergency rations. Now, however, the world's gourmets are looking forward to being able to buy some of the deep-sea fish that their old recipe books mentioned, such as halibut, haddock, plaice and tuna.

There are far fewer poisons around than there used to be. The agreement and enforcement of national laws and global treaties have seen to that, supported by the diligent activism of millions of citizens. Whistle-blowers, using mobile phone web connections to share direct observations with activists, journalists and enforcement agencies, make cheating almost impossible. The cost of safe waste disposal now has to be included in the annual business plans and accounts of all companies and cities, and the concept of 'safe disposal' has long since ceased to include dumping at sea. A result is that those sea fish that are available may be expensive, but they are at least safe to eat. Some businesses and towns went bankrupt when these laws were first put into effect, but environmental reconstruction grants and other forms of targeted public assistance smoothed the path for most. A clear legal framework for investment, and iron deadlines set well in advance, also helped the private sector make the necessary adjustments. Many companies were aided by their customers, who deliberately chose only to buy the products of those who joined the new standards early on.

But the most strategic change has been the stripping of greenhouse gases from the open air by vast thermonuclear-powered scrubbers, the liquefied carbon dioxide and stabilised methane hydrates being stored by the gigatonne deep underground in the long-empty sea bed oil and gas fields. These mechanisms are the main sources of profit for the heirs of the major energy companies, and are paid for by taxes on carbon emissions. They've proved even more effective than the signatories of the 2022 post-Kyoto IV treaty had imagined. Greenhouse gas concentrations in the atmosphere peaked at 0.05 per cent carbon dioxide equivalent in 2052, their growth having been slowed somewhat by earlier desperate measures to reduce emissions, but are now down to 0.04 per cent and falling. The oceans stored a lot of energy in those few dangerous years in an enhanced greenhouse, and sea levels are still rising, rainfall patterns remain distorted, and wild storms still pound coastal areas. But there is a sense of hope nevertheless. Soft engineering, new building codes, and the relocation of settlements and populations has allowed for considerable adaptation, and most people are now reasonably safe.

Tourism is still important, having recovered a bit from the almost-complete closure of commercial aviation during the mid-century famines. Now, high-speed air travel is used only for emergencies, weather permitting, and the usual way to travel long distances is by train. The new tourism is mostly a positive force for everyone. Coastal communities treat visitors rather like fish, as a renewable resource for sustainable harvesting by locals. Almost every community in the world is in direct contact with every other through the web, and environmental education has been a big thing everywhere for decades. Thus standards are extremely high at the grass roots, and there is no question anywhere of shoddy or damaging tourism investments being allowed through ignorance of proper planning or equity standards. Corruption and incompetence have been minimised through direct local elections, the active participation of citizens' groups in government, and the free flow of information.

Many resort areas that were ruined in earlier waves of tourism development, where they haven't had to be abandoned to the sea and seaborne storms, have long since been rehabilitated, building on their unique strengths, the capacity of nature to regrow, and the friendliness of local people who no longer fear for their livelihoods. In any case, most visitors want to be ecotourists and are happy to pay what it takes for a high-quality holiday, without damaging the world too much by travelling around in it. The greenest of them, of course, stay at home and tend their gardens instead. Or else they do voluntary ecosystem restoration work, driven by the need somehow to honour and atone for the deaths of tens of millions of species in the recent mass extinction.

And what of fresh water in all this? The key change came with the mass movements of mid-century, in which the ideas of water democracy became widespread. As locally accountable management of ecosystems became the norm, and communities learned from one another about what to require of their leaders, these ideas came to be expressed in a host of different ways, grafted onto a range of religions and philosophies of life. They were represented by phrases such as 'Water is nature's gift, essential to all life, and connecting all life', and 'Water must be free for sustenance use, but is limited and must be conserved', and 'No one has a right to abuse, waste or pollute water or water-bearing ecosystems'. In the same package came the strong sense of duty to use water sparingly, caringly and justly, and also the belief that all people and all species have a right to their necessary share of water. Many also concluded that water is unique and by nature a common resource, so it can't be owned as private property or sold as a commodity. In short, out of hardship, necessity, wisdom and local power, water became sacred again.

The practical results were incredibly diverse, as people sought ways to solve their own local water crises, each resulting from a different local interaction among ecosystems, climate, weather, terrain, drainage, proximity to the sea, population, wealth and culture. And this was the whole point of local people seeking and gaining the power to make their own choices, putting their own water ethic into practice in their own way. They did this by restoring floodplains and welcoming floods, guiding rainwater down wells and into deep tanks, and building check-dams, all to recharge ground water. They did it by harvesting dew, building sea water greenhouses and other structures to condense water vapour from the air, using solar desalination, and catching and storing rain in domestic and community cisterns. They did it by banning thirsty crops, and seeking out species and cultivars that use the least water. They did it by rediscovering ancient ways, such as underground tunnels, to harvest water from aquifers at sustainable rates. And they did it by finding old or new ways, such as bamboo or ceramic feeders, to deliver water drop by drop to the roots of growing crops, rather than smearing it across waterlogged and salinising fields. All had their places in community water strategies around the world.

Meanwhile, other configurations of cities, citizens, catchments, rivers and ground waters came into play. City administrations learned to strike long-term deals with catchment dwellers to pay a fair price for catchment services, in return for their help in protecting upstream ecosystems. Cities learned to collaborate with one another to liberate rivers from industrial canals and dams, so that they and their shadows, and migrating fish, could run free again. They also learned to collaborate with landowners to encourage organic farming and low-impact use of ecosystems in the catchments. Their citizens, also affected by the ideas of water democracy, quickly became adept at conserving municipal waters, and began insisting that public water companies (the private ones having been repossessed) fix all leaks in their supply systems. They also began to demand that their urban environments become much more porous, with parks and gardens everywhere, all roads made of water-permeable materials, and a water trap on every roof. Finally, in a development that decisively changed the lives of consumers and producers around the world, sophisticated labelling and rationing arrangements for the use of virtual water came to be accepted, along with similar means to limit the use of embedded carbon.

By now, the attitude that gave rise to this new deal for city dwellers and water is fixed in the popular culture as the 'New York state of mind'. This is because New York City provided one of the earliest and clearest examples of an ecological service in action. It had long taken much of its water from forests and reservoirs in the Catskill Mountains, carried by the 190 km Catskill aqueduct. As the city developed back in the twentieth century, pollution threatened its drinking water. Faced with a choice between an expensive filtration plant and environmental restoration in the Catskills, the city chose restoration. Using an environmental bond issue to raise cash, it bought up land in and around the catchment and provided incentives for sustainable resource management. The New York state of mind is also, of course, famously 'can do', fiercely individual yet also strongly communal. These are all attributes that are valued highly in 2085.

PATHS TO PROGRESS

Did you find this vision of the future appalling, or appealing? Regardless, something along these lines is now necessary, whether we like it or not. The question is really how we get there with the least possible hardship for people and damage to nature. In this, everyone is a key player. With a global culture using the biosphere as a global resource, everyone is a participant now. Can it be done? Well, we humans may be ferocious conquerors of nature, but we can also be good negotiators and brilliant at solving problems. There is no law that says we have to use ecosystems to death. With the same skills that we have shown in the past, we can invent ways to use limited resources more and more efficiently, even sustainably.

Where does this leave us? With an almighty set of problems, a fairly clear idea of where we want to go, and a lot of specific things to do to get there. People have managed with far less over the last 100,000 years. It is often said that the world has shrunk into a global village. Perhaps a smallish city is more accurate, with richer and poorer suburbs, industrial quarters, museums, parks, pavements, thoroughfares and patches of urban blight. In the past, when communities were threatened by environmental collapse, many were able to save them-selves. They did this by paying attention, by co-operating, by negotiating, and by choosing wise and accountable leaders. We still have these skills, and they will still work. They must.

BLASTS FROM OUR PASTS

What can ordinary people do about the world's water crisis? To answer that we should appreciate what we've *already* done to change important things, and consider how we did it, and what we can learn from it. For we have slain many of the monsters of the past, and we've invented new and better ways to settle disputes, to negotiate win-win outcomes, and to generate wealth all round, and we've done it through cultural change, with a determination to get freedom and votes, and an equal determination to use them. We've learned how to make a fuss in order to get what we want in our own lives. As our culture has evolved, in no particular direction but generally towards greater convenience and safety, we've managed to obtain, or are in the process of obtaining, or realise we'd like to obtain, such things as decent schools, effective medicines, wholesome food, clean air, and fresh water without sewage or poison in it. I'd like to tell a few stories that illustrate how we got some of these benefits, to show what kind of struggles and tactics have proved necessary to success in the past. All will shed some light on the kinds of things that need to be done by ordinary people and extraordinary ones (if there is a difference) to sort out the water problem as a whole, and the tens of thousands of water problems that make it up.

The wholesome food chain

From early in the twentieth century, petrol-driven tractors and other machines transformed the energetics of farming. Fields grew larger and cropping more intense to make more efficient use of this machinery. Cheap nitrogen-based fertilisers allowed farmers to lose interest in natural soil fertility. Then, from the 1950s onward, there were further dramatic advances in mechanisation, including large-scale irrigation, as well as the mass production of pesticides, starting with DDT. Plant breeders responded by creating crop varieties that did well in irrigated, fertilised and simplified ecosystems. The resulting package of cultivars, chemicals and technologies was rolled out around the world in the form of a 'Green Revolution' that massively increased the production of staple crops. All of this had predictable impacts on landscapes, communities, water use, soils and biodiversity, many of them devastating. The process has continued through the growth of major corporations that are using new genetic techniques to modify organisms, for instance to make crop plants resistant to herbicides, thus allowing poisons to be used even more indiscriminately against 'weeds'. This is the farming industry responsible for the bulk of food now consumed worldwide.

Organic foods, as advertised, are much less likely to contain pesticide residues, which anyone in their right mind would prefer them not to do. And, more to the point for an ecologist writing about water, they are much more likely to have been grown in a diverse ecosystem by someone who cares about soil microbes, the balance of nature and clean water. Indeed, in order to qualify for the 'organic' label, that farmer must demonstrably be running such a farm. And if that's the case, then that farm would not be leaking pesticides and fertilisers into the ground water, it would not be eroding and losing its fertility while silting up and over-fertilising rivers and lakes, and it would be sustaining many species of wild plants and animals in addition to the varieties it's growing for sale. So, an organic farm is contributing solutions to the world's crises of ecosystem destruction, water supply and contamination, and mass extinction. What I don't understand is why any other kind of farming is legal.

Ah, but we won't go there . . . instead, our culture requires that we rely on 'consumer choice'. This means piling vaguely toxic and not-very-tasty produce next to 'organic' produce, and slapping a huge price tag on the latter, creating a tax on common sense and environmental virtue. Or, looked at another way, unless taxes, subsidies or legal restraints work against doing so, it's much cheaper to make nasty food in simple ecosystems than it is to make wholesome food in complex ones. Consumers can then choose to take advantage of the lower prices, thus helping to destroy the living world and poison their families.

It is up to us to generate change, supported by education and clear, assured product labelling. Since the early 1990s, the retail market for organic produce in developed countries has been growing by about 20 per cent each year, despite the price differential. Simply put, we increasingly don't trust conventional producers to offer us safe food, or governments to regulate them properly, and we're developing a taste for better food that also helps save the world. And this is happening despite the lack of official subsidies to help level the playing field in favour of organics.

The sanitation contagion

The reek of urban England is eye-watering, clawing at nose and throat, retch-making. The fast-growing cities of this modernising country are awash in the overflow and leakage of cesspits. The Cray, Wandle, Lea, Ravensbourne and Thames in London, the Tame, Rea and Cole in Birmingham, the Irwell, Medlock, Irk and Mersey in Manchester – all these rivers are open sewers, sluggish and bubbling in the sun. In the back streets, hidden in corners or on bundles of soaked rags, the sick and the dead decay in the summer heat. Flies are everywhere. It is 1832, women have just been denied the vote in the first Reform Act, and 7,000 people have just died of cholera. But neither the great stink nor the deaths of the poor are passing features. They are permanent parts of urban industrial life.

Within five years, though, the Office of the Registrar General had been formed, to begin registering births and deaths. It promptly started to produce a stream of the names of the dead that began to generate public concern during the first years of Queen Victoria's reign. And five years later, in 1842, Edwin Chadwick's great report on *The Sanitary Condition of the Labouring Population of Great Britain* caused a storm. It said, in short, that people were dying like flies because clean water offered by private companies was unaffordable, and that ordinary people were spending their lives amidst foul smells and festering sewage, which made them ill. His chief recommendations were for public institutions to take responsibility for water supplies and sewage disposal, and for there to be clean water and a sewer connection in every home.

Another five years passed before parliamentary time was spent on these matters, but then came the Public Health Act of 1848 and the Metropolitan Water Act of 1852, which mandated the public provision of clean water. Added impetus came with further cholera outbreaks, and then with the conclusion reached by John Snow in 1854 that cholera was a waterborne infection of some kind. This he deduced by observing the cholera epidemic of 1853, which killed 12,000 Londoners, and tracing the pattern of victims to a single water pump on Broad Street in the Golden Square area. But Snow's theory was not immediately accepted by the medical establishment, which was dominated by those, such as Chadwick, who believed that miasmas and befouled waters caused diseases, not germs. Thus it was not the 'contagionist' group, those who believed in microbes, that had most influence on public sanitation during the rest of the nineteenth century, but the 'anticontagionists', who held that 'sewer gas', 'bad air' and 'filth' were the enemies of health.

The anticontagionists committed themselves to a massive project against filth. Parliament was forced to close by the stench of the Thames in the hot summer of 1858, its members retiring to their country estates. The poor had no such escape, and continued to die. By 1880, municipalities had replaced private companies as the main providers of water in towns and cities. But although average income doubled, and life expectancy increased slightly, child mortality remained stubbornly high. Children carried on dying, mainly of diarrhoea and dysentery, at much the same rate as they would still be doing in tropical developing countries more than a century later. By the end of the 1870s, it had become clear that public water supply was only part of the solution, and the streets and rivers still ran with excrement. As the anticontagionists would have put it, people were still exposed to foul sewage airs.

There was mounting political pressure for public action, and sanitation became a rallying point for social reformers, municipal leaders and public health bodies. The national élite increasingly saw poor sanitation as not just disgusting and sinful, but as a real constraint on economic prosperity. A surge in public investment followed in the newly industrial, and now imperial, Britain. This was often financed in new ways to avoid higher taxes, with cities supplementing low-interest loans from central government through municipal borrowing on bond markets. By the end of the nineteenth century, a quarter of local government debt was due to water and sanitation expenditure. Capital spending per person on sanitation rose more than four-fold between 1885 and 1905. And at last child mortality began to fall, and life expectancy to rise, both steeply. Between 1900 and 1910, infant mortality fell by nearly 40 per cent, from 160 to 100 deaths per 1,000 live births. Thus the cannon fodder of the First World War were saved for their fate.

While we're being cynical, it's worth looking at why there was such a time-lag between the public acquisition of water supply in 1860–80 and public investment in sanitation in 1885–1905. One answer is that the water reforms were driven by the owners of new industries, who needed cheap, public water for their factories and labourers. The sanitation reforms, by contrast, were driven by the Third Reform Act of 1884, which extended voting rights to the poor.

Comparable events were unfolding across the Atlantic in much the same period, during which the urban population of the USA exploded into the tens of millions, swept repeatedly by diseases. Key ideas on what to do about this reached the USA through Chadwick's 1842 report, which inspired Lemuel Shattuck to write America's first comprehensive public health plan in 1850. Like Chadwick, he adhered to the miasma theory of disease, yet his recommendations on waste disposal, pollution and preventative medicine might have been written with bacteria in mind. No fewer than 36 of his 50 recommendations were still standard public-health practice a century later.

The 'sanitary movement' in America faced two problems, both of them familiar from Victorian England and still deadly to people in the tropical slums of the world today. These were that early water systems had been improved by private companies, so poor households could not afford to be connected to them, and that those systems let sewers empty back into the water supply. In England, public water for the poor was provided first, followed 25 years later by sanitation; in America things happened the other way round. Sanitation campaigns started first, in the 1850s, but the extension of public water supplies only began in 1900, and New Orleans, where African Americans were dying from typhoid at roughly twice the rate of whites, municipalised its water only in 1908. This reversal may reflect Americans' greater enthusiasm for private enterprise in the form of water companies, and greater tolerance of hardship among the poor.

The early start on sanitation responded to the fact that mid-nineteenth-century American cities were death traps of filth. A survey of New York City revealed sewage, and blood and offal from slaughterhouses, running in the streets among the overcrowded tenements. This led to a public outcry and the creation of a Metropolitan Board of Health, dedicated to sanitary reform. Chicago, meanwhile, had been enduring repeated outbreaks of cholera and dysentery, and in 1855 a new Chicago Board of Sewage Commissioners appointed Ellis Sylvester Chesbrough to design a proper sewage system. The aim was to drain wastes into the Chicago River and Lake Michigan, but for this to happen the whole downtown area had to be rebuilt three metres higher than before. Chesbrough achieved this, with some public health gains, but drinking water was still coming from the polluted lake. That problem was solved in 1900 by building the Chicago Sanitary and Ship Canal, which made the Chicago River flow backwards into the Illinois River, thus polluting the Mississippi instead.

The sanitary crusade spread across the USA, led by George E. Waring, another anticontagionist and an advocate of flush toilets and the sanitary reform of entire communities as a way of getting rid of bad air. Waring made his fortune by designing sewage systems based on Chadwick's ideas, starting in Memphis after a yellow fever outbreak in 1878. He promoted the theory and design concepts nationwide, and eventually became New York City's most effective Commissioner of Street Cleaning. With exquisite irony, he died of yellow fever in 1898, in Cuba, while trying to fight an outbreak in Havana by cleaning up the city's bad air problem, never knowing that yellow fever was caused by a mosquito-borne virus rather than by sewer gas.

The miasma theory may have been wrong, but its believers nevertheless built a sanitary infrastructure that hugely improved public health, probably more than all the later inventions of modern medicine combined. Further improvements were needed, though, and the contagionists promoted measures specifically aimed against bacteria, such as water filtration and chlorination. By 1940, these had reached half of all Americans, and they played a key role in boosting life expectancy at birth by sixteen years from 1900 to 1940, in slashing child mortality, and in almost eliminating typhoid fever. Every life saved in this way cost about US$500 in 2002 prices, but every dollar spent generated another US$23 in increased output and reduced health costs, giving a high economic return on public investments in sanitation.

Cleaning the air

In early December 1962, I was sitting on the floor of the upstairs bedroom I shared with my brother, at home in south-east London. I remember the window at the far end of the room being open. Maybe I'd opened it to have a closer look at the yellow-brown blanket that was pressed against the pane. It was daylight, or should've been. The yellow-brown stuff oozed through the gap at the bottom of the window and drifted to the floor. It stank. It was a foul mixture of water vapour condensed as droplets around billions of tiny particles of carbon and tar, mixed with sulphur dioxide that was partly dissolved in the water to make sulphuric acid. Over the next few days, 750 Londoners would die of it, and thousands more would suffer chest pains, inflamed lungs, emphysema and permanent lung damage. What I was looking at was the last of the great London smogs, caused by a temperature inversion that trapped cold air and fog near the ground, while at the same time hundreds of thousands of homes were burning a sulphur-rich coal called 'nutty slack' to keep warm.

London had been famous for smoky fogs since Victorian times, but when over 4,000 died in a few days of smog in 1952, and another 8,000 in the weeks afterwards, people began to ask questions. The science and medicine had come a long way by then, but not quite far enough, it seems, as the smog deaths were at first attributed to a flu outbreak. When the same thing happened again in 1955, though, the Minister for Housing and Local Government, Lord Duncan-Sandys, introduced the Clean Air Act of 1956. My mother, who remembers the time well, says that this was very much a personal initiative of the Minister, and Sir Terry Farrell, writing in the *Independent* fifty years later, described it as 'a brave example of political leadership, by . . . a great conservationist, ahead of his time, who got the Act through, in spite of opposition from much of [the] cabinet.'

MAKING SUSTAINABLE MUNICIPALITIES

Other cities have brought air pollution under control, including Seoul in South Korea, and Tokyo and Kitakyushu in Japan, and have improved waste management and water supply systems. It has been said that a threshold in per-person income must be reached before any clean-up becomes possible. The idea is that individual wealth brings with it more opportunities to learn about and campaign on health and environmental issues, and also a greater collective ability to pay, through taxes and markets, for environmental solutions. There are other mechanisms, though, which are not necessarily wealth-dependent. These include people becoming aware enough, motivated enough and organised enough to encourage municipal governments to clean up the environment, and private corporations to stop polluting it. Even very poor communities can take effective action once the oppressive effects of environmental deterioration are recognised, especially if they enjoy strong and accountable leadership.

Several cities show what can be done with a combination of imaginative municipal leadership and access to knowledge of what can be accomplished. Curitiba in south-eastern Brazil, for example, had massive problems of unemployment, slums, pollution and congestion. Twinning Curitiba with Hangzhou, one of the most beautiful cities in China, the Curitiba administration overcame their environmental problems by a mixture of means. Investment in a clean public transport system slashed air pollution, and the system is now used by 1.3 million people daily. A programme of waste separation and recycling grew until two-thirds of the city's daily waste was being processed. Then, a soft engineering approach to flooding and recreational space led to the creation of 2,100 hectares of porous parks, woods, gardens and other open spaces, mostly along river banks and in valley bottoms, where they act as water flow regulators during the rainy season. Curitiba, and its population of 1.6 million, was awarded the United Nations' highest environmental prize in 1990 by the UN Environment Programme.

Meanwhile, on the other side of the world, the 6.2 million-strong city of Dalian in north-east China was one of the most heavily developed and polluted industrial areas of that country. It established a twinning arrangement with Kitakyushu in Japan, which had already made spectacular progress against pollution. The arrangement allowed for the training of factory managers, the refitting of factories, and the development of a local government environmental zone. Environmental improvements in Dalian since 1990 led to its municipal government being elected to UNEP's Global 500 Roll of Honour for outstanding contributions to the protection of the environment.

While the iron's hot

In 1986, the Marcos dictatorship in the Philippines was overthrown in a 'people power' revolution. The next few years were a time of intense reform, during which Aquilino Pimentel became Minister of Local Government and was then elected to the Senate. He led the preparation of the Local Government Code, which was made law by Congress in 1991. This new law transferred many powers from central government to the various levels of local government – the provinces, municipalities and barangays. These were made responsible for most services in agriculture, public works, social welfare and health, as well as for community-based forestry projects up to 50 km² in area and the enforcement of fisheries and environmental laws. The local governments were given increased tax-raising powers and a 40 per cent share both of the national tax base and of revenues from using local natural resources. Finally, the code also strongly encouraged NGOs to take an active role in developing local autonomy. These changes were particularly effective in giving the authority for defensive environmental decision-making, such as the protection of water catchments, to the 1,554 municipalities of the country – electorates of a few thousand people represented by mayors and powerful local councils working closely with social and environmental NGOs.

Since then, every election has brought more 'green' mayors to power, and local environmental initiatives have multiplied – a community water catchment forest here, a marine fish sanctuary there, spreading fast as people learn that they work. As people have become used to these new responsibilities, they have become more reluctant to accept dodgy schemes devised by distant élites. The impact of the Local Government Code was overwhelmingly liberating and enormously beneficial for the environment. It allowed the effects of decades of top-down environmental exploitation to be reversed, as local people followed their instincts in safeguarding the ecosystems that provide them with water supplies, fish and security against flashfloods and landslides. This law was in the spirit of the times, but the spirit quickly changed in Congress, if not in the countryside. Attempts to repeal it soon began, as national élites who had been displaced with the fall of Marcos recovered their influence. From 1992 to 1998 Senator Pimentel was attacked by all manner of the dirty tricks for which politics in the Philippines is famous. But both he and his law survived, and he was re-elected to the Senate in 1998.

WHAT HAVE WE LEARNED?

Farming and forests

From the organic farming story, we learned that we can get informed, then choose what to buy and what not to, and that corporations will respond. This isn't a power to be taken lightly. Corporations do not have morals or values that can be appealed to, but the customer is still king and has to be responded to, or corporations collapse. Other examples of our using our purchasing power include the certification of wood and wood products by the Forest Stewardship Council. Driven by public interest in stopping deforestation, this was set up in 1993 to develop standards for the sustainable management of forests and the labelling of their products. By 2003, a third of Dutch consumers were able to recognise the FSC logo; by 2005, the value of FSC-labelled products exceeded US$5 billion worldwide; and by 2006, a total of over 68 million hectares had been FSC certified. By then, non-FSC timber products were becoming hard to find in many shops and construction projects in the UK and elsewhere. Also in 2006, Random House became the first big publishing group to be FSC certified. Then in 2007, the ING Bank committed itself to using FSC-certified products, as did Warner Music in many of its CD and DVD products. Meanwhile, both Bloomsbury and Scholastic, the publishers of the final *Harry Potter* book, agreed to use FSC paper in printing it, the largest contract ever issued for supplying such paper for a single print run.

Fairtrade and fisheries

Somewhat similar is the spectacular growth of Fairtrade certification, a system that since 1997 has let people identify products that meet agreed environmental and social standards. It is overseen by a standard-setting body, FLO International, and a certification body, FLO-CERT, and involves independent auditing of producers to ensure that standards are met and maintained. By the end of 2006, 569 producer organisations in 58 developing countries were certified. During 2006, Fairtrade-certified sales were about €1.6 billion (US$2 billion) worldwide, a 41 per cent increase over 2005, and an estimated 1.5 million disadvantaged producers were benefiting directly, and another 5 million indirectly. This represents a small, but quickly growing share of world trade, that benefits a small, but quickly growing share of the world's population.

The Marine Stewardship Council was also established in 1997, to identify and certify the best-managed fisheries, and label their products so that people could choose to buy them. By 2007, its assessment process was underway for fisheries that accounted for 42 per cent of the global wild salmon catch, 32 per cent of the global prime whitefish catch (cod, pollock, hake, haddock, ling and saithe), and 18 per cent of the global spiny lobster catch. By then, 22 fisheries had been certified, and 608 of their products labelled for sale in the world's shops. Changes like this are driven partly by the visionary efforts of individuals, but are sustained through *our* choices about what we want to pay for, choices that businesses then respond to. We live by buying things from corporations, and every buying decision we make sends a signal to the corporate world.

Bravery under fire

From decentralisation in the Philippines, we learn that, even under prolonged dictatorship, public opinion will grow in favour of local control of environments, to which brave leaders may sometimes be able to respond. But we also learn that such windows of opportunity may be brief, so changes need to be implemented fast and in ways that create a level of public support that makes them irreversible. Then, from the cleaning of London's air, we learn that chronic, harmful pollution will eventually be identified as such, that public opinion will gradually shift in favour of controlling it, and that conditions allowing brave and effective political leadership will finally occur. And from the re-birth of Curitiba and Dalian, we learn that the leaders of cities can help their citizens transform urban environments, often helped by sharing ideas with other cities through twinning arrangements. These principles apply equally to the challenge of preserving or restoring water quality, and protecting water-bearing ecosystems.

Water and sanitation

Finally, we can learn several things from the story of water and sanitation in England and America. First, that filthy environments *can* be transformed, and households that are starved of clean water *can* be relieved. But first people have to know there's a problem, and that they can fix it. Too often people accept prevailing conditions as normal and unchangeable, and even filthy air can become accepted as something that goes with the territory of urban life. Several cities, including London, have been known affectionately as the 'Big Smoke' in their time. I once visited a friend's house in an exclusive part of Colombo, Sri Lanka, and he apologised for the appalling, lurid quality of the water in the canals around the private estate. I replied, 'Lal, you and your neighbours are among the most educated and prosperous people in this country – you could fix it if you wanted to. The question is: why don't you want to?' After huffing and puffing for a bit, Lal admitted that he'd never thought of it quite like that.

Dedicated and energetic campaigners are often needed, to articulate and form public opinion, to make propaganda, and to exert influence on those who make decisions and laws. Beyond a certain level of understanding, the motivations and knowledge base of those campaigners can vary greatly, without necessarily affecting the impact of their enthusiasm. This we saw from the competing miasma (or filth) and germ (or contagion) theories, with their similar practical implications for public health reform. One might say the same about a drive to save a water-bearing ecosystem, with some reformers doing so because it's beautiful (or sacred) and others because it's functional (or cheaper than the alternatives).

On the other hand, different campaigners may have different interests, leading to different outcomes, or outcomes achieved at different times. The variables of public versus private water supply, and more versus less public investment in sanitation, play out very differently according to the relative influence of rich and poor when decisions are being made. We have also seen that economic calculations that give a high value to avoided social costs can be used to rationalise public, though probably not private, investment in water and sanitation.

Finally, water supply and sanitation improvements in England and the USA in the mid-nineteenth to the mid-twentieth centuries were done locally, city by city. Central government may have mandated certain things (such as the public supply of water and the separation of sewage), subsidised things (such as through grants and low-interest loans) and authorised things (such as municipal debt), especially in England, but it was up to the city authorities in each case to act locally.

Lessons of the divided mind

Through all this we see operating the two opposed themes of the human mind, which earlier in the book I described as 'Taoist' and 'Confucian'. The first of these kinds of thinking can be described as more liberal, flexible and holistic, more suited to ecology. The second can be seen as more imperial, mechanistic and reductionist, better suited to engineering. Without wanting to make too much of the labels, I think we can see 'Confucian' thinking behind the concentration of power among élites, the decisions that favour their interests, the corporate monopolisation of water supplies and the exclusion of the poor from their benefits, the reaction against local empowerment, and the great engineering schemes such as the filtration plant offered to New York City as an alternative to maintaining its ecological relationship with the Catskill Mountains. And I think we can see 'Taoist' thinking behind the opposite trends, those towards learning, sharing, empowering and acting locally to achieve harmony with nature.

The future I sketched at the start of this chapter could only possibly be achieved through a balancing of these two approaches, with both needed but in different ways and places. Roles, rights and responsibilities for sustainable and just outcomes must be agreed, but then the settlements will need to be protected by the vigilance of citizens, and sometimes enforced by draconian penalties. The finding of thousands of different solutions to tens of thousands of different water crises by millions of different communities is 'Taoist' in flavour, but they must all comply with the sometimes-harsh rules of ecology. And then, some of the greatest strategic challenges can only be overcome through the use of awesomely 'Confucian' technologies, like the satellites, the fast police boats, and the industrial plants needed to clean the atmosphere in a global emergency.

WHAT SHOULD WE DO NOW?

Preparing for a better future

While gathering our strength for the challenges ahead, we can do much to prepare. We can pay attention to the real world and its governments, agencies, charities and corporations that claim to act in our name, or that offer us things that they want us to buy, often on monopoly terms. We can help build public understanding of the value of ecosystems and water catchment services, and strengthen local people's ability to defend them. We can create and share knowledge about the economic and social value of catchment ecosystems, and of how to solve institutional problems that get in the way of sensible solutions. We can encourage and reward investigative journalists for exposing unjust and unsustainable arrangements. We can ensure that water is priced effectively to include the true cost of ecosystem maintenance, and that the cost is spread fairly to take account of the rights of all people to a living amount of clean, fresh water. And meanwhile, we can take charge of our buying and voting decisions to send consistent, powerful signals to the élites of our world, telling them that we can no longer be taken for granted, and that wise use of ecosystems is now mandatory.

What we do depends on who we are and where we live, but a top priority is always going to be to understand where the water we use comes from, and at what social and environmental cost. Is someone else, or some distant ecosystem, being deprived of water so that it can run freely out of our taps? Is the money we pay for water being used to restore and maintain the water catchments and distribution system, and if not, why not? Or is it being wasted on leaking pipes and excessive profits for water companies? How proof against climate change is the whole arrangement? It may work fine today, but has someone really got their head round the implications of longer hot seasons, or salt intrusion, or a change in weather from weeks of soaking rain to brief intense storms, or the extra demands being placed by a sudden growth in new housing?

Some suggestions for things that we can do to protect the future were included in previous chapters. For *oceans*, we can support precautionary fishing, for example by asking for seafood certified by the Marine Stewardship Council. We can be more pro-active in making sure that our visits to coasts help to reward sustainable, local initiatives, such as reef-guarding. For *wetlands* and *rivers*, we can encourage public debate about wetland and floodplain management, and take an interest in planning decisions that may increase everyone's vulnerability. For *lakes*, we can help tighten understanding of the implications of different kinds of conflicting uses, and encourage dialogue and planning among different stake-holders. For *ground waters*, we can insist that our cities are as porous as possible, that rainfall is used to recharge aquifers, and that ground water extraction is limited to natural recharge rates. And for *all waters*, we can agitate for labelling to tell us the virtual water content of everything we buy. As with the advice in Chapter 8 to find out if a water charity knows what it's doing before you help fund it to drill boreholes, there's no substitute for finding out all you can about every issue raised in this book, and encouraging the flow of knowledge to other consumers and to decision-makers.

To do today

Even the smallest things make a difference, like putting a water butt in the garden to collect the rain. We can refuse to buy bottled water, until it's in bottles with deposits on them, and a share of the price goes to ensuring safe water for all. Come to that, we can demand that the tap water we pay for is safe, clean and pleasant to drink, rather than buying a filter and letting the water companies off the hook, and we can ask those water companies what exactly they're doing about water quality and conservation.

More altruistically, we could give each other really useful presents, like a village-scale safe-water unit from Oxfam (www.oxfamunwrapped.com), or sign the petition for a universal right to water at www.watertreaty.org, or we could find out about what other charities are doing, guided by what's in this book, identify some of the best and give to them. Or, going more global, how about having a look at the United Nations' water activities (www.unwater.org), and helping to celebrate World Water Day on 22 March each year? Or, more locally, how about cleaning up a stream – why not join or found a local volunteer group to do just that?

Eventually, we'll have to do much more than this, by responding to every opportunity to advance towards a sustainable world, one that doesn't die during the twenty-first century. The ideas of water democracy offer us a good place to start building our thoughts and determination to organise and to act. No one should ever think that personal action can't solve problems, no matter how huge and remote they may seem. We just can't tell whether or how something we do will stimulate events in a positive way. By setting an example, who knows what will happen? We can send multi-media messages to corporations and politicians. We can become 'citizen scientists' and help monitor environmental change. We can find ways to commemorate the million species that we are killing each year, and promise to do everything we can to stop the mass extinction. We can get informed, stay informed, and use our knowledge. We can buy less, buy local and buy green. Indeed, this is perhaps our greatest power, which we can start using immediately. From this moment we could, if we wanted, buy *only* organics, *only* fairtrade, *only* sustainably certified seafood, wood and paper. And we could go on from there. Knowledge is power. Pass it on.

APPENDIX 1. GLOSSARY

Anthropocene. 'The age of mankind', a name for the current geological age, reflecting the profound influence that humanity is having on nature.

APELL. Awareness and Preparedness for Emergencies at Local Level, the United Nations' disaster preparedness training process.

Aquatic. Living in and around water.

Aquifer. A 'water bearer', an underground layer of wet, porous, rocky material.

Atom. The basic unit of matter. Each has a nucleus, made of positively charged protons and uncharged neutrons, and an orbiting cloud of negatively charged electrons.

Atmosphere. (1) The envelope of gases around a planet. Earth's atmosphere has a well-mixed troposphere up to between 7 km (near the poles) and 17 km (near the equator), with a stratosphere above that to about 50 km, and other layers still higher. It is overall about 78 per cent nitrogen, 21 per cent oxygen and 1 per cent argon, along with water vapour and trace gases such as carbon dioxide. (2) An atmosphere is also a unit of pressure, being the average atmospheric pressure at sea level, or 101,325 pascals (i.e. about 100 kilopascals or kPa).

Biodiversity. The variety and variation among all kinds of life, or the information that has accumulated in living systems, including the genetic coding for proteins, metabolic pathways, cells and individual organisms, the differences among lineages and species, and the relationships and processes in every ecosystem.

Biology. The study of how organisms live, reproduce and evolve.

Biomolecule. A molecule involved in the metabolic chemistry or structure of an organism.

Biosphere. All parts of the Earth where life occurs, comprising the atmosphere, oceans, fresh waters, soils, and their underlying sediments and rock layers.

By-catch. The organisms that are caught and killed by accident during fishing operations, currently about a quarter of the 120 million tonnes of marine wildlife harvested each year.

Cambrian. A period of geological time, between 542 and 488 million years ago, which is the first to show abundant fossils of complex, multi-cellular organisms. Time between the formation of Earth (about 4.6 billion years ago) to the start of the Cambrian is known as the pre-Cambrian (comprising the Hadean, Archaean and Proterozoic eons).

Cambrian Explosion. An evolutionary event, between 530 and 520 million years ago, when all the basic patterns of modern life forms originated.

Carrying capacity. The idea that for any species in an ecosystem, numbers are limited by the resources produced by that ecosystem under prevailing conditions.

Catchment. An area where all water flows to a common destination, bounded by a line of terrain where water flows elsewhere.

Coastal. Lands close to the sea and shallow waters close to the land.

Compound. Joined atoms of more than one element.

Confucianism. A philosophy founded by Confucius, a Chinese scholar who lived from 551 to 479 BC, whose teachings deeply influenced Chinese and other societies in East Asia and elsewhere. It values meritocratic rule and filial piety, with every individual knowing their place in the social order and acting according to that place. When used as a state religion or ideology, it is associated with authoritarianism, paternalism and submission to authority.

Covalent bond. A bond between two atoms caused by electrons from each being attracted by the nucleus of the other, so that the electrons become located in the space between the two atomic nuclei. Once there, the electrons and nuclei are pulled together, but the two positively charged nuclei also repel each other, so the atoms stay apart at a distance where the attractive force balances the repulsive force.

Desertification. A process usually involving a combination of over-grazing and poor farming practices that expose the soil, drought that weakens it, and wind that blows it away. This destroys plant communities and soils, and degrades the landscape to a point where it looks superficially like a natural desert.

DIPECHO. The disaster preparedness training process of the European Commission's humanitarian aid department.

DNA. Deoxyribonucleic acid, a complex biomolecule that contains coded information or instructions for making RNA, and thereby proteins, in the cells of organisms.

Dowsing. The use of a dowsing or divining rod to search for underground water.

Ecology. The study of how organisms live together, meet their needs for energy and nutrients, and respond to opportunities and challenges in their environment.

Ecosystem. All the organisms living in a place and time, all the relationships among them, all the physical features of light, heat, moisture, wind, waves and chemistry that affect them, and the history of the place as well.

EEZ. Exclusive economic zone, an area of the ocean where a nation claims the exclusive right to fishing and sea-bed mining.

Element. A basic kind of matter. Every atom belonging to one element has the same number of protons in its nucleus.

There are 94 elements that occur naturally on Earth (plus another 24 or so that have been made in nuclear reactors).

El Niño-Southern Oscillation. A global cycle that links oceans and the atmosphere, and is the most prominent known source of variation between years in rainfall around the world. It has characteristic influences in the Pacific, Atlantic and Indian Ocean basins, and in surrounding countries.

Endemic. A species or higher taxon, like a genus or a family, which occurs in the wild nowhere other than in a particular place, and which has not done so in the historical past.

Feedback. The changing of a process or system by its own results or effects. Negative feedback suppresses the cause of change, such as where increased numbers of predators decline in numbers because they have reduced the abundance of their prey. Positive feedback amplifies the cause of change, such as where warm weather melts ice, which reduces reflected sunlight and increases warmth.

FSC. Forest Stewardship Council, a body that sets standards for certifying wood and timber products as having come from a sustainably managed forest.

Gaia. The ancient Greek Goddess who personified the Earth, now used to mean a complex entity involving the biosphere, atmosphere, oceans and soil, all of them being parts of a feedback system that maintains conditions favourable to life.

Greenhouse gas. A gas that is transparent to visible light but less so to heat energy. In the atmosphere, greenhouse gases, such as water vapour, carbon dioxide and methane, trap sunlight as heat, thus contributing to the greenhouse effect and increased global warming.

Hard engineering. The use of physical structures to oppose natural forces, for example massive concrete sea walls to resist waves.

Hydration. The process of absorbing or combining with water.

Hydration shell. The shape of water molecules created around a polar molecule or ion, like a mould or negative image of a substance that water has encountered.

Hydrogen bond. Bonds between molecules that occur when one molecule has a peripheral electron pair or a hydrogen atom, and the other molecule has the same. The positively charged hydrogen atoms of one molecule are attracted to the negatively charged electron pairs of the other, and vice versa. Hydrogen bonds are about ten times stronger than the forces of attraction between polar molecules, but about ten times weaker than the covalent bonds between atoms.

Hydrophilic. A polar molecule that has an attraction to water.

Hydrophobic. A non-polar molecule that is repelled by water.

Hydrosphere. That part of the Earth where water is found, including the biosphere and deep parts of the planet that are too hot for life.

Hydrous minerals. Minerals that incorporate water within their molecules, formed when surface waters seep underground, dissolve and react with various solids, and are then heated, pressurised, cooled and dried over time. Examples are rock salt, gypsum, opal, olivine, serpentine and kaolin. Hydrous minerals formed at higher temperatures include marbles, micas and quartzites such as amethyst.

Ion. An atom with a net charge, due to the loss of an electron because of the impact of radiation, or because of the approach of a strongly charged nucleus to a more weakly charged one.

Ionic bond. A bond between two atoms caused by attraction between positive and negative ions.

Isotope. An atom that contains a different number of neutrons than other atoms of the same element.

Light-year. The distance travelled through a vacuum by light in a standard Julian year of 365.25 days, or 9.461 trillion kilometres.

Mangrove. A type of tropical, coastal swamp forest made of trees that can grow in salty, tidally flooded mud.

Mass extinction. A brief period when many species and lineages were lost almost simultaneously from the fossil record, events that are known from 488, 444, 350, 251, 200 and 65 million years ago.

Metabolism. All the chemical reactions that occur in living cells.

Methanogenesis. A kind of cellular respiration that uses carbon rather than oxygen to catch electrons after they have been used to produce the energy needed for life. The carbon can come from any number of small molecules (such as carbon dioxide and acetic acid), and an exhaust gas is methane (CH_4). Methanogenesis is used by archaeans and other microbes, and is the last step in the decay of organic matter. Much methane is produced by rotting vegetation where there is little oxygen available, for instance in swamps, dam lakes and flooded rice fields, and through the complex biochemistry inside the guts of mammals, where microbes digest cellulose and other plant materials. Averaged over a century, methane is about 25 times more potent as a greenhouse gas than carbon dioxide.

Molecule. A structure of two or more atoms bonded together.

MSC. Marine Stewardship Council, a body that sets standards for certifying fish and other marine products as having come from a sustainably managed fishery.

MSY. 'Maximum sustainable yield', the theoretical point at which animals can be taken from a wildlife population at the greatest possible rate without causing it to decline.

Natural selection. The process by which environmental factors cause differences in breeding success among similar organisms.

NGO. Non-governmental organisation, usually a non-profit, public-interest group that is registered as a charity.

Organic. Relating to or derived from living matter, or, in chemistry, a molecule containing carbon atoms and associated with life.

Organism. A living thing, such as a plant, animal or fungus, or a microbe such as a bacterium, archaean or protist.

Photosynthesis. The use of energy in light to support metabolism, mainly the use of sunlight by plants and other organisms to induce carbon dioxide and water molecules to combine, thus creating sugars, with oxygen as a waste product.

Polar molecule. A molecule with one part, side or end that is positively charged and another that is negatively charged.

Precautionary fishing. Fishing in which catch rates and techniques have been proven not to harm fish populations and their supportive ecosystems.

Proteins. Large biomolecules, made of amino acids, that are essential parts of organisms and participate in every process within living cells and organisms.

RNA. Ribonucleic acid, a complex biomolecule that contains coded information or instructions for making proteins in the cells of organisms.

Shadow river. Water that has soaked into a river bed and is slowly following a similar course towards the sea (also known as the hyporheic flow).

Soft engineering. The use of living systems and their natural resilience to oppose or modify natural forces, for example by planting mangrove forests to absorb the impact of waves.

Stakeholder. Someone with something to gain or lose in a potential dispute over resources.

Storm surge. The destructive arrival on land of a volume of water that has been raised by the low atmospheric pressure at the heart of a storm.

Sustainable. Able to be maintained, and therefore, of an ecosystem, able to continue more-or-less unchanged despite being used by people.

Symbiosis. An intimate relationship involving mutual dependence and shared adaptation by the partners (in biology between organisms, metaphorically between water and life).

Taoism. A philosophy, originating in China around 300 BC, that takes its name from the Tao, which is the flow of the universe and the influence that keeps it balanced and ordered. It sees a flow of life energy (*Qi*) in the body and in nature, and regards the whole Earth as being alive with it. Thus, Taoism is about living with the flow of life energy in the world, and focuses on naturalness, vitality, peace, spontaneity, humanism, detachment, the strength of softness, and passivity or 'effortless doing'. Taoism values intuitive wisdom over rational knowledge, spontaneous action over planned action, and harmony with nature over domination of nature.

Taxon. A group of related organisms, such as a species, a genus (a group of related species), a family (a group of related genera) and an order (a group of related families).

Troposphere. The inner layer of the atmosphere, from ground level to about 20 km high at the equator, which is constantly mixed by rising air, but at its top gives way to the jet streams and the stratosphere above them.

TURF. 'Territorial use rights in fisheries', asserted by communities wishing to make their own decisions on how to manage their environment and fish stocks.

UNEP. United Nations Environment Programme.

Virtual water. The amount of water used to make things, for example a tonne for a kilo of wheat, 3 t/kg for sugar, 5 t/kg for rice, 20 t/kg for coffee, and 24 t/kg for beef.

Wetland. Land ecosystems strongly influenced by water, and aquatic ecosystems with special features due to shallowness and closeness to land. These include swamps and marshes, lakes and rivers, wet grasslands and peatlands, oases, estuaries, deltas and tidal flats, near-shore marine areas, mangroves and coral reefs, and human-made sites such as fishponds, rice paddies, reservoirs and salt pans. The definition may be simplified as ground that for natural reasons is tidally, seasonally or occasionally under water or, if always under water, where it's shallow enough for ground-rooted vegetation to grow up through.

WFD. The European Commission's Water Framework Directive (2000), which requires integrated river basin management, and aims to ensure clean rivers and lakes, ground water and coastal beaches throughout the European Union.

APPENDIX 2. READING LIST

Here are some of the publications I used in writing this book. Other information came from my own field work, magazines such as *Oryx* and *New Scientist*, and from the Internet, especially Wikipedia, with multiple sources being used to confirm one another.

ActionAid, *Unjust Waters: Climate change, flooding and the protection of poor urban communities – experiences from six African cities* (ActionAid, London, nd).

Philip Ball, *H₂O: A Biography of Water* (Weidenfeld & Nicolson, London, 1999).

Margaret Barber and Gráinne Ryder (eds.), *Damming the Three Gorges: What dam builders don't want you to know* (Earthscan, London and Toronto, 1993).

John D. Barrow, *The Artful Universe* (Penguin Books, London, 1995).

David Blackbourn, *The Conquest of Nature: Water, landscape and the making of modern Germany* (Jonathan Cape, London, 2006).

Fritjof Capra, *The Tao of Physics* (Wildwood House, London, 1975).

—— *The Web of Life* (HarperCollins, London, 1996).

Charles Clover, *The End of the Line* (Ebury Press, London, 2004).

Elizabeth C. Economy, *The River Runs Black: The environmental challenge to China's future* (Cornell University Press, Ithaca, 2004).

GESAMP (IMO/FAO/UNESCO-IOC/WMO/WHO/IAEA/UN/ UNEP Joint Group of Experts on the Scientific Aspects of Marine Environmental Protection) and Advisory Committee on Protection of the Sea, *A Sea of Troubles* (Rep. Stud. GESAMP No. 70, 35 pp, 2001) Written by Geoffrey Lean.

R.E. Johannes, *Words of the Lagoon: Fishing and marine lore in the Palau District of Micronesia* (California University Press, Berkeley & London, 1981).

William Bryant Logan, *Dirt: The ecstatic skin of the Earth* (Norton & Co., New York, 1995).

James Lovelock, *The Revenge of Gaia* (Penguin/Allen Lane, London, 2006).

Mark Lynas, *Six Degrees* (Fourth Estate, London, 2007).

Bernadette McDonald and Douglas Jehl (eds.), *Whose Water Is It? The unquenchable thirst of a water-hungry world* (National Geographic, Washington DC, 2003).

George Monbiot, *Heat* (Penguin/Allen Lane, London, 2006).

Elaine Morgan, *The Scars of Evolution* (Souvenir Press, London, 1990).

—— *The Aquatic Ape Hypothesis* (Souvenir Press, London, 1997).

Fred Pearce, *When The Rivers Run Dry: What happens when our water runs out?* (Eden Project Books, London, 2006).

C. Pye-Smith and G.B. Feyerabend, *The Wealth of Communities: Stories of success in local environmental management* (Earthscan, London, 1994).

Kirsten Schuyt and Luke Brander, *Living Waters: Conserving the source of life* (WWF and the Free University of Amsterdam, 2004).

Vandana Shiva, *Water Wars: Privatization, pollution, and profit* (Pluto Press, London, 2002).

United Nations Development Programme, *Beyond Scarcity: Power, poverty and the global water crisis*, 2006 Human

Development Report (Palgrave Macmillan, New York, 2007).

United Nations Environment Programme, *Freshwater in Europe: Facts, Figures and Maps* (Division of Early Warning and Assessment, Geneva, 2004).

—— *One Planet Many People: Atlas of our changing environment* (Division of Early Warning and Assessment, Nairobi, 2005).

—— *In the Front Line: Shoreline protection and other ecosystem services from mangroves and coral reefs* (World Conservation Monitoring Centre, Cambridge, 2006).

—— *Environmental Degradation Triggering Tensions and Conflict in Sudan* (Post-Conflict and Disaster Management Branch, Geneva and Nairobi, 2007).

—— *2007 Global Environmental Outlook* (Division of Early Warning and Assessment, Nairobi, 2007).

Diane Raines Ward, *Water Wars: Drought, flood, folly, and the politics of thirst* (Riverhead Books, New York, 2002).

Waterwise, *Hidden Waters, A Briefing* (Waterwise, London, 2007).

E.O. Wilson, *Biophilia* (Harvard University Press, Cambridge, Massachusetts, 1986).

—— *Consilience: The Unity of Knowledge* (Knopf, New York, 1998).

—— *The Future of Life* (Knopf, New York, 2002).

—— *The Creation: An Appeal to Save Life on Earth* (Norton, New York, 2006).

World Resources Institute, *World Resources 2005: The wealth of the poor – managing ecosystems to fight poverty* (WRI, Washington DC, 2005).

—— *Millennium Ecosystem Assessment: Ecosystems and human well-being – wetlands and water synthesis* (WRI, Washington DC, 2005).

ACKNOWLEDGEMENTS

Mary Monro is a total star for reading every word at least twice, improving most of them, and deleting the rest. I'm also grateful to Diane Banks, my energetic agent; Elaine Morgan, who provided inspiration and corrections for the section on aquatic ape theory; Tony Durham and Sue Bishop at ActionAid, who shared insights and experience on water and poverty; Roger Hammond at Living Earth, who scattered bright ideas freely; and Ed Faulkner and Davina Russell at Virgin Books, who read and guided wisely.

INDEX